Robert Stawell Ball

MECHANICS

Elibron Classics
www.elibron.com

Elibron Classics series.

ISBN 0-543-86728-5 (paperback)
ISBN 0-543-86727-7 (hardcover)

This Elibron Classics Replica Edition is an unabridged facsimile
of the edition published in 1879 by Longmans, Green, and Co., London.

THE LONDON SCIENCE CLASS-BOOKS

ELEMENTARY SERIES

EDITED BY

PROF. G. C. FOSTER, F.R.S. AND PHILIP MAGNUS, B.Sc. B.A.

MECHANICS

MECHANICS

BY

ROBERT STAWELL BALL, LL.D., F.R.S.

ANDREWS PROFESSOR OF ASTRONOMY IN THE UNIVERSITY OF DUBLIN
ROYAL ASTRONOMER OF IRELAND

LONDON
LONGMANS, GREEN, AND CO.
1879

LONDON: PRINTED BY
SPOTTISWOODE AND CO., NEW-STREET SQUARE
AND PARLIAMENT STREET

EDITORS' PREFACE.

NOTWITHSTANDING the large number of scientific works which have been published within the last few years, it is very generally acknowledged by those who are practically engaged in Education, whether as Teachers or as Examiners, that there is still a want of Books adapted for school purposes upon several important branches of Science. The present SERIES will aim at supplying this deficiency. The works comprised in the SERIES will all be composed with special reference to their use in school-teaching; but, at the same time, particular attention will be given to making the information contained in them trustworthy and accurate, and to presenting it in such a way that it may serve as a basis for more advanced study.

In conformity with the special object of the Series, the attempt will be made in all cases to bring out the educational value which properly belongs to the study of any branch of Science, by not merely treating of its acquired results, but by explaining as fully as possible the nature of the methods of inquiry and reasoning by which these results have been obtained. Consequently, although the treatment of each subject will be strictly elementary, the fundamental facts will be stated and discussed with the fulness needed to place their scientific significance in a clear light, and to show the relation in which they stand to the general conclusions of Science.

In order to ensure the efficient carrying-out of the general scheme indicated above, the Editors have endeavoured to obtain the co-operation, as Authors of the several treatises, of men who combine special knowledge of the subjects on which they write with practical experience in Teaching.

The volumes of the Series will be published, if possible, at a uniform price of 1*s.* 6*d.* It is intended that eventually each of the chief branches of Science shall be represented by one or more volumes.

G. C. F.
P. M.

AUTHOR'S PREFACE.

In the earlier chapters of this book I have occasionally adopted the methods used by Delaunay in his well known 'Cours Elémentaire de Mécanique.' Not a few of the illustrations have been derived from the same source.

The question of units naturally demands a few words of explanation. I am fully sensible of the great advantages of the method of expressing forces in absolute measure, but still I have not thought it wise to adopt that method. My experience in teaching mechanics to beginners is decidedly against it. Mechanics cannot be taught without constant reference to physical illustrations and to ordinary machines and appliances. Even if the teacher commences with the determination to recognise no force but so many *dynes* or *poundals* (or whatever the unit may be called), he can hardly be quite consistent. The temptation to speak of the force overcome by a pulley-block, for example, as so many pounds or tons will occasionally prove irresistible; and a sort of double standard is set up which is certainly most undesirable. I have therefore adopted the gravitation of one pound as the unit of force, and I have endeavoured to use this unit consistently.

ROBERT S. BALL.

February, 1879.

CONTENTS.

CHAPTER I.

PRELIMINARY NOTIONS ON MOVEMENT.

CHAPTER II.

PRELIMINARY NOTIONS ON FORCE.

CHAPTER III.

THE PRINCIPLE OF THE LEVER.

CHAPTER IV.

EQUILIBRIUM OF PARALLEL FORCES.

CHAPTER V.

THE PARALLELOGRAM OF FORCE.

CHAPTER VI.

GRAVITATION.

CHAPTER VII.

ILLUSTRATIONS OF THE EQUILIBRIUM OF FORCES.

CHAPTER VIII.

THE INCLINED PLANE.

CHAPTER IX.

FRICTION.

CHAPTER X.

ENERGY.

CHAPTER XI.

PRACTICAL ILLUSTRATIONS OF MECHANICAL PRINCIPLES.

CHAPTER XII.

THE MOTION OF A FALLING BODY.

CHAPTER XIII.

THE SECOND LAW OF MOTION.

CHAPTER XIV.

THE THIRD LAW OF MOTION.

MECHANICS.

CHAPTER I.

PRELIMINARY NOTIONS ON MOVEMENT.

§ 1. **Introduction.**—It will be found, as we proceed in the study of Mechanics, that this science is intimately connected with the varied phenomena known generally by the term *movement*. Our ordinary experience presents us with a thousand different forms of movement, and obviates the necessity of attempting any rigorous definition of that term. The movements of our own bodies and those of other animals, of vehicles and machinery, of the clouds and the heavenly bodies, are sufficient illustrations of the kind of phenomena to which we refer. It may perhaps be thought that the objects we have mentioned are so exceedingly diverse in their character, and that the nature of the movements which they execute are so complicated and so various, that it would be impossible to find any common feature which pervades them all. What, for example, is the common feature in the movement of a cloud and the movement of a cricket-ball? Yet they have a common feature, for each of them is in *the act of changing the position which it occupies.*

When once it has been recognised that there is a common feature in all cases of movement, the first step has been taken towards understanding the classification of different kinds of movement. It remains to consider some of the different ways in which it is possible for an object to change its position.

§ 2. **Uniform Motion.**—Take first a very simple example. Imagine a level and straight railway, along which a train is travelling, and let us also add the supposition that the train is moving uniformly. By moving uniformly we mean that the train is steadily maintaining a rate of so many miles per hour or so many yards per second. In this case, regarding the engine and the entire train as a single object, we have the most simple case of motion which can be conceived, i.e. the *uniform movement of an object in a straight line.*

If the railway be not straight, but have curves and inclines (as in fact all railways have), then the motion of the train will be *curvilinear.* If, further, the speed of the train be uniform, we have the case of the *uniform movement of an object in a curved line.*

In the case of the uniform movement of an object either in a straight line or a curved line the distance travelled is doubled when the time is doubled, or is tripled when the time is tripled, or in general the *distances accomplished are proportional to the times which have been employed.* It is thus easy to solve such problems as the following, relating to uniform motion.

1. Being given the distance travelled in a specified interval of time (for example, a mile in a minute), to find the distance passed over in any other specified interval of time (for example, an hour).

2. Being given the distance travelled in a specified interval of time (for example, four miles in an hour), to find the time which would be necessary to accomplish any other specified distance (for example, twenty miles).

In the first case it is obvious that if an object moves over a mile in a minute it will be able to accomplish sixty miles in an hour, provided the same rate of travelling is maintained.

In the second case, if the object moves at the rate of four miles per hour it will accomplish twenty miles in a period of five hours.

By an application of the ordinary rule of three we are enabled to solve many other questions which may be proposed in which uniform motion is considered. For example, a steamer travels 720 miles in three days, how much will it accomplish in a week? The answer is readily found to be 1,680 miles. Or conversely, we should find that a steamer which required seven days to move through 1,680 miles would traverse 720 miles in three days.

§ 3. **Velocity.**—The *velocity* of a body moving uniformly is *the distance which the body traverses in an interval of time which is chosen arbitrarily as the unit.* For example, we may say that a railway train is moving with a velocity of sixty miles an hour, or of a mile a minute.

If we know the distance which the body travels in

a given time when moving uniformly, then we can find the velocity with which the body is travelling by dividing the distance travelled by the time the journey has occupied. So also, when we know the velocity with which the body is moving, it is easy to calculate either the distance which can be accomplished in a given time, or the time which will be required to travel a given distance.

§ 4. **Examples.**—Adopting the second as the unit of time, then a body which moves over a distance of 9,630 feet in one minute and a half has a velocity of $9{,}630 \div 90 = 107$ feet per second.

Conversely, a body which is moving with the uniform velocity of 107 feet per second will in one minute and a half accomplish a distance equal to $107 \times 90 = 9{,}630$ feet; or if the distance were given, then the time which would be required would be $9{,}630 \div 107 = 90$ seconds.

If the velocity be twelve yards per minute, what space will be accomplished in three hours fifteen minutes? We solve this problem by multiplying the given time in minutes into the velocity; and the answer is therefore $195 \times 12 = 2{,}340$ yards.

The explosion of a cannon is heard seven seconds after the flash is seen, and the distance of the cannon is known to be 7,700 feet; determine the velocity with which sound travels. We see at once that the sound travels in one second over a distance of $7{,}700 \div 7 = 1{,}100$ feet, which is, therefore, the required answer. It should be remarked that we have here supposed that the light from the flash travels with such velocity that the time it occupies is inappreciable.

§ 5. **Ordinary Units.**—The units of length and time most commonly employed in England are the *foot* and the *second.* There are, however, several other units in occasional use, but these can be all ultimately reduced to the foot and the second. The units actually chosen depend upon the magnitudes of the quantities to be estimated, the degree of precision which is required, and the actual processes employed in the measurement.

Usage has established several customs which cannot but be recognised. Thus, it is the custom in navigation to express the rate at which a vessel is moving as so many 'knots': for example, a vessel which is going eight knots is moving at a rate which, if maintained uniformly for an hour, would carry the vessel over eight nautical miles.

§ 6. **Uniform Movement of Rotation.**—When a wheel is rotating round its axis, the number of revolutions accomplished in a given time expresses the rapidity with which it is turning. Let A B D (fig. 1) be the circumference of the wheel, which is revolving about the fixed centre C, and let the radius A C be the direction of one of the spokes of the wheel. As the wheel revolves, the radius A C is, of course, carried with the wheel, and is thus moved into positions B C, D C, which are inclined to the original position by gradually increasing angles. We can thus say that one fourth of a complete revolution of the wheel is equivalent to an angular displacement of one right angle, or

Fig. 1.

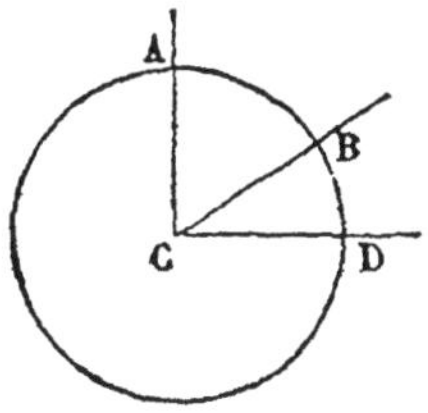

of 90° ; half a revolution is an angular displacement of 180°, and a whole revolution is an angular displacement of 360°.

The number of degrees through which the wheel has turned, or the number of complete revolutions which it has accomplished, will express clearly the amount of angular displacement which the wheel has undergone. *The movement of rotation is said to be uniform when the angular displacement is proportional to the time, whatever the duration of that time may be.*

A simple illustration will explain the necessity of the last clause. The second-hand of a clock accomplishes one complete revolution in one minute, two in two minutes, three in three minutes, and so on. It thus appears that the number of whole revolutions made is strictly proportional to the time, but the motion is still not uniform, for the second-hand moves by jerks, and the angles described in very short intervals of time are not proportional to the times in which they are executed.

§ 7. **Angular Velocity.**—The angular velocity of a uniform movement of rotation is the angular displacement which is accomplished in that specified interval of time which is taken as the unit. We can thus express the angular velocity in various ways; for example, we may take the number of degrees moved through in a second, or the number of complete revolutions which are made in a minute.

In the majority of ordinary cases, such as the toothed wheels used in machinery and the fly-wheels of steam-engines, the angular velocity is usually expressed by the number of revolutions accomplished in a minute. When the angular velocity is expressed in this way, it is easy to find the total number of revolutions accomplished in a given time; for it is only necessary to reduce the given time to minutes and to multiply the result by the number expressing the velocity.

The various questions that may arise with respect to uniform angular velocity involve three quantities, viz., 1st, a specified interval of time; 2nd, the total number of revolutions accomplished in that time; 3rd, the number of revolutions accomplished in one minute. When any two of these quantities are known, then the third can be found by ordinary arithmetic.

Suppose a wheel rotating uniformly makes ten revolutions per minute; then in an hour it will have made 600 revolutions, and it will require ten minutes to accomplish 100 revolutions. It will require one-tenth of a minute, or six seconds, to accomplish one revolution, and in one second it will have accomplished one-sixth of a revolution. In one second the angular displacement of the wheel will be 60°, and this is, therefore, the angle through which each radius of the wheel turns in one second.

§ 8. **Velocity of a point in a revolving body.**—When a wheel is revolving, every point in the circumference of the wheel describes a circle of which the centre of the wheel is the centre. So, in general, whenever a body is revolving around a fixed axis, every point in the body describes a circle of which the centre lies in the fixed axis. As we can calculate the length of the circumference of the circle by the ordinary rule, we are able to ascertain the actual distance through which a point in the body moves. If, therefore, we know the number of revolutions which the body makes in a minute, we are enabled to ascertain the actual distance traversed by each part of the body in a minute, and therefore the velocity of the point becomes known.

§ 9. **Illustration.**—The principles to which we have here referred are very well illustrated by a useful mechanical appliance known as the *endless saw* (fig. 2). This machine consists essentially of two drums, over which passes a flat and flexible steel band, of which one edge is notched to form a saw. One of the two drums is turned round by a steam-engine, or some other source of power. This causes one side of the saw to move upwards and the other to move downwards. Against the descending portion of the saw, the piece of wood is pressed which is to be cut.

FIG. 2.

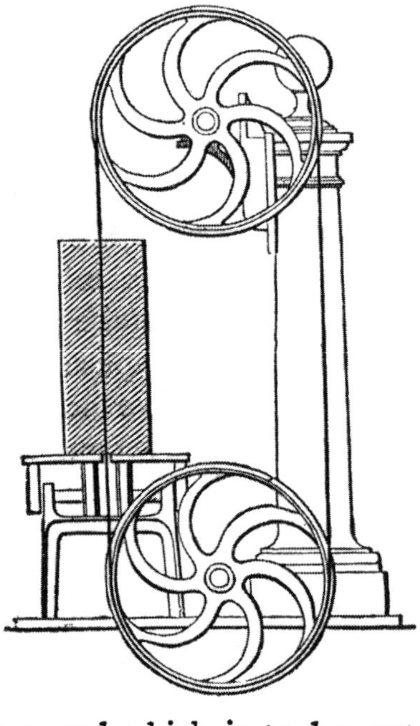

To calculate the velocity of the endless saw we shall assume that the circumference of the driving drum is one yard and that it makes fifty revolutions per minute.

In one minute a point on the circumference of

the drum will move over a distance of $3 \times 50 = 150$ feet; and therefore the velocity is 150 feet per minute, or 2·5 feet per second.

§ 10. **Geometrical Representation of Uniform Motion.**—It is often convenient to represent the motion of a point in a graphic manner. Draw two straight lines O X, O Y (fig. 3), intersecting each other at right angles in the point O.

FIG. 3.

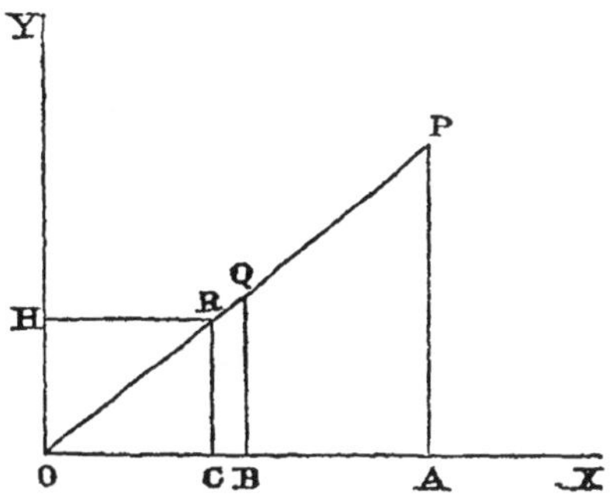

We may adopt any convenient scale for expressing the distances and times. Let O A denote the distance traversed in the time which is represented by the line A P, drawn parallel to O Y, and therefore perpendicular to O X. Join O P, then the time occupied in describing *any* given distance, O B, can be readily found by drawing B Q parallel to O Y, and applying our adopted scale to determine the time equivalent to B Q.

For the triangles O B Q and O A P are similar, and therefore

$$OB : OA :: BQ : AP;$$

hence it follows that the condition characteristic of uniform motion, namely, that the distance is proportional to the time, is fulfilled.

The same construction will enable us easily to determine the space which will be described in a given time, for mark off on O Y a distance O H, which represents the given time, and draw through H a line, H R, parallel to O X. From R let fall R C perpendicular upon O X, then the distance O C is obviously the required length.

It is thus easy to determine the velocity, for if the distance O H be taken as the unit of time, then O C will be the distance described in the unit of time, i.e. the velocity.

It will thus be seen that the inclination of the line O P to O X depends upon the velocity of the motion. The smaller that inclination the greater the velocity, and *vice versâ*.

EXAMPLES.

1. If a railway train moves 10 yards per second, how far will it travel in half an hour? *Ans.*: 10 miles 400 yards.

2. If a horse travels 9 miles in 57 minutes, what is his velocity in feet per second? *Ans.*: 13·9.

3. The diameter of a circus is 10 yards, and a horse goes round it twice in a minute, what distance will he accomplish in a quarter of an hour? *Ans.*: 2827·4 feet.

4. Two stations are 9 miles apart. A train starts at the same time from each station at the rate of 15 and 25 miles per hour respectively. How soon will they meet and how far will the quickest train have travelled? *Ans.*: 13 min., 30 sec.; 5·625 miles.

5. The fly-wheel of an engine makes four revolutions per minute, its circumference is 50 feet. Determine the distance travelled by a point on its circumference in one second? *Ans.*: 3 feet, 4 inches.

CHAPTER II.

PRELIMINARY NOTIONS ON FORCE.

§ 11. **The First Law of Motion.**—What is known in mechanics as the first law of motion is thus stated.

'*Every body continues in its state of rest, or of uniform motion in a straight line, except in so far as it may be compelled by impressed forces to change that state.*'

No one who is competent to judge of the matter can doubt for a moment that this law is perfectly true, but it cannot be proved in the same sense as the propositions in Euclid are proved. The first law of motion has been shown to be true by induction, and without entering upon the question as to whether it is necessarily true we can assert that, taking the universe as we find it, this law invariably prevails. It has been tested in every way that we can suggest, and always with the same result. It may, therefore, be received without the slightest hesitation as a fundamental law of nature.

There is one part of this law which is, indeed, sufficiently obvious. No one can doubt that a body which is at rest will remain for ever at rest unless some cause intervene to set the body in motion. A stone on the ground will remain there indefinitely unless by the influence of some external cause it be displaced. An apparent exception might be found in the case of a living animal, which seems to have the power of setting itself in motion. This is, however, no more a real exception than is a locomotive engine, or a mechanical toy, which runs along the floor under the action of a wound-up spring. In order that an animal may set itself in motion, it has to press against the ground or some other fixed object; the result is, that the ground presses back with exactly equal force against the animal, and it is this reacting force which really causes the animal to move. Similarly, it is the reaction of the rails or of the floor respectively which enables the locomotive or the toy to move. Such cases, however, will be better understood when the Third Law of Motion has been discussed.

There is, however, one part of the first law of motion which at a first glance is apt to seem at variance with common experience. If a body were once set moving, the law asserts that the body would continue to move uniformly in a straight line for ever, *provided no impressed force act upon the body.* The reason why this appears to be at variance with experience is that we are unable practically to realise the condition which it imposes, of placing a body under such conditions as not to be acted upon by any forces. But we can, however, convince ourselves of the truth of this statement by the following reasoning: A stone thrown along the road soon comes to rest. The stone leaves the hand from which it is thrown with a certain velocity and receives no more force from the hand. Hence, if no other force acted upon the stone, we should expect to find (if the first law of motion be true), that the stone would continue to run on for ever with the velocity it had at the moment of leaving the hand. But other forces do act, the attraction of the earth pulls the stone down, and when it begins to bound and roll upon the ground, a new force called the force of friction begins to act. Friction is due to the roughness of the road, and it very soon brings the stone to rest. But let the stone be thrown upon a surface of smooth ice, then when the stone is sliding upon the ice, the friction which resists the motion is comparatively small, though it still exists, and the stone will run on to a very great distance. We thus see that the stone does not stop *itself*, but that it comes to rest in consequence of the friction of the surface on which it moves. The first law of motion amounts to the assertion that, if it were possible to withdraw the stone completely from the action of friction and all other external forces, it would run on indefinitely in a straight line with unabated velocity.

It therefore appears that to impart motion to a quiescent body, or to cause a moving body to change the direction in which it is moving, or the velocity with which that motion is executed, the presence of a certain cause is necessary. Whatever this cause may be, it is called force. There are several different descriptions of force, of which we shall mention a few.

FIG. 4.

§ 12. **Gravity.**—If a stone be let fall from the hand, it falls to the ground. The force which acts upon the stone is called *gravity*. All the objects which surround us are acted upon by gravity. It is gravity which causes the descent of rain from the clouds and the motion of the water in a river.

If a body be suspended by a string (fig. 4), the string hangs in the direction in which the body would fall if allowed to do so. This is therefore the direction along which the force of gravity acts. If we alter the nature of the body or its weight or bulk, the direction of the string remains unaltered; it is always perpendicular to the surface of water at rest, and is called the *vertical.*

A line perpendicular to a vertical line is called a *horizontal line.* A plane perpendicular to a vertical line is called a *horizontal plane.* A plane which contains a vertical line is called a *vertical plane.*

§ 13. **Molecular Forces.**—When an elastic body —a piece of india-rubber for example—is deformed without passing the limit of elasticity, the body when released tends to resume its original form. In this case the particles of the body move in consequence of certain forces which act upon them, tending either to draw them farther apart or closer together as the case may be. The forces which produce these movements of the particles of a body are termed *molecular forces.* It is to molecular forces we are indebted for the power which drives our steam-engines. Under the influence of heat the particles of the water repel each other, so that the water assumes the much more bulky form of steam. The molecular forces by which this expansion is effected are so powerful that the steam in expanding is able to exert a great degree of pressure, which is turned into a useful form by the steam-engine.

§ 14. **Electrical Attractions.**—A different class of forces are presented in the phenomena of *electricity.*

A piece of sealing-wax rubbed with silk will attract pieces of paper, or other small objects, in virtue of the electricity with which it becomes charged. Closely allied to these electrical forces is the well-known *magnetic force* by which an ordinary magnet attracts a piece of iron towards it.

§ 15. **Pressures and Tensions.** — A force may sometimes be applied to a body without producing any movement in that body. For example, a stone may be lying at rest upon the table, but the force of gravity is acting upon the stone, for if the table were suddenly removed the stone would immediately fall. So also a stone may be at rest when suspended by a cord from a fixed point, but gravity is still in action, for if the string were cut the stone would, of course, fall. Whenever force does not produce motion it gives rise to a *pressure* or a *tension*. Thus in the first case the stone exerts *pressure* upon the table which supports it, and in the second case it produces a *tension* on the cord by which it is suspended.

§ 16. **Force manifested by Changing Motion.**— The existence of force may also be made manifest by the alteration or the destruction of motion which it can produce. Suppose a ball to be thrown vertically upwards. While the ball was in the hand it was acted upon by the force which the arm exerted, but when the ball leaves the hand it is no longer acted upon by the arm, but ascends merely in consequence of the motion which has been communicated to it. Here the arm has exerted force and produced motion and the ball ascends; but the force of gravity is constantly endeavouring to move the ball in the opposite direction, and as the ball receives no reinforcement of its upward motion, the never-ceasing gravity gradually robs the ball of its movement, until the moment comes when, having attained its greatest height, the ball is for an instant at rest. But the force of gravity still continuing to act, the ball descends with gradually increasing velocity until it reaches the ground. Here, during the ascent of the ball, gravity was acting to destroy motion, and during the descent acting to produce it.

§ 17. **Inertia.**—The facts that a body cannot *by itself* pass from a state of rest into a state of motion, or from one state of motion into a different state of motion, are often expressed by saying that matter is *inert*, or that it possesses *inertia*. The following illustrations are for the purpose of showing some of the different phenomena arising from inertia.

A man is riding on horseback, and the horse stops suddenly. If the horse has previously been going rapidly his rider is in great danger of being thrown over the horse's head. Why is this?

The man has a tendency to continue moving in the same direction as that in which he was moving before the horse stopped, and if he be not able to resist this tendency, he will be carried forwards over the horse's head.

In starting a railway-train, the speed of the train increases *gradually* until full speed is attained, but why could not full speed be obtained at once?

In the first second the power of the engine is expended in giving the train a certain very small velocity. At the commencement of the next second, the train has, therefore, already acquired some velocity and is disposed to continue moving with this velocity even if the engine were removed. But as the engine is still attached, the train receives another increase of speed during the second second, so that it enters on the third second with an increased velocity and so on.

By increasing the power of the engine, the time required to get up full speed may be diminished, but under all circumstances the operation must be a gradual one.

It may, however, be asked what is the limit to the increase of speed which may be attained. The limit is due to the resistances to which the motion of the train is exposed. The resistances partly arise from the retarding action of the air, and the effect of this resistance *rapidly increases with the velocity of the train.* When by the increase of the velocity the resistances have become equal to the drawing-power of the engine, no more augmentation of speed is possible. We may in such a case consider the power of the engine as being entirely spent in overcoming the resistances, and the train may then be regarded as a body moving along with a uniform velocity without being acted upon by any forces.

Why is it dangerous to alight from a railway-carriage before the train has stopped? Because your body participates in the motion of the railway-carriage. When, therefore, your feet are brought to rest by reaching the ground you are apt to be thrown down in the direction towards which the train is going. The more rapidly the train is moving the greater will be the danger of this happening.

It is stated that railway collisions might often be

prevented were the trains supplied with adequate brake-power. What is the explanation of this? Assume for simplicity that the line is quite level, and that a train is going at full speed. The engine-driver sees, let us suppose, a few trucks standing on the line in front of him, and he becomes aware that a collision is imminent unless his train can be stopped. His train, being a body in motion, can only be stopped by the application of force; and what forces has the engine-driver at his disposal? If he merely turns off the steam, then the ordinary forces of resistance will gradually diminish the velocity of the train, and finally bring it to rest; but this will not be prompt enough to avert the danger. He may also reverse his engine, which will, no doubt, accelerate the effect of the ordinary resistances in bringing the train to rest; but even this will not be adequate. He requires the aid of some other powerful force. Later on, we shall consider with more detail the very important force of friction, which resists the sliding motion of one surface over another. By forcibly pressing a brake against the wheels, they can be prevented from turning round and made to rub along the line, whereby a great force of friction (§ 60) is evoked which brings the train speedily to rest. The larger the number of wheels which can be thus prevented from turning, the more efficiently can those in charge of the train control its movements.

§ 18. **Motion of a Projectile.**—A stone thrown into the air in any other than the vertical direction, describes a curve, and then falls down to the ground. If it were not for the action of gravity the stone would continue to move on indefinitely in the direction in which it has been thrown, but gravity is continually acting upon the stone, and therefore deflects it more and more from the direction in which the motion began. Thus the curved path of the stone is explained.

Questions.

1. State the first law of motion.

2. If a ball be thrown along the ground it soon stops. If the ball be thrown into the air it moves in a curvilinear path. Are these facts consistent with the first law of motion?

3. A heavy metallic top set spinning in the air will not remain spinning nearly so long as it would do in a vacuum. Can you explain this?

4. A man standing up in a boat is liable to be thrown down if the boat is suddenly stopped by running against the bank. Why is this?

5. If gravity acts upon all bodies and tends to draw them towards the earth, how is it that a balloon will ascend?

CHAPTER III.

THE PRINCIPLE OF THE LEVER.

§ 19. **Measurement of Force.**—We shall find it necessary to establish a definite method by which the magnitudes of forces may be estimated. Take any standard weight—for example, a pound—then the force with which this weight presses upon the table on which it stands, or the force which must be exerted in order to hold this weight in the hand, may be taken as the standard of force. We may for brevity speak of this as a *force of one pound*. A force of two pounds, three pounds, &c., will denote the pressure produced by a weight of two pounds, three pounds, &c., and in general the magnitude of a force may be expressed by the number of pounds to which it is equivalent.

You will gradually learn that all forces can really be expressed in this way, even though the sources of these forces be very different from the attraction of gravitation which is the cause of the pressure which is exerted by a weight placed upon the table.

§ 20. **Direction of a Force.**—If a heavy body be pulled along the table by a string, the body moves in the direction of the string, and this is said to be the direction of the force exerted by the string upon the body. In order to ascertain what will be the effect of a force, it is necessary to know the direction in which it acts.

FIG. 5.

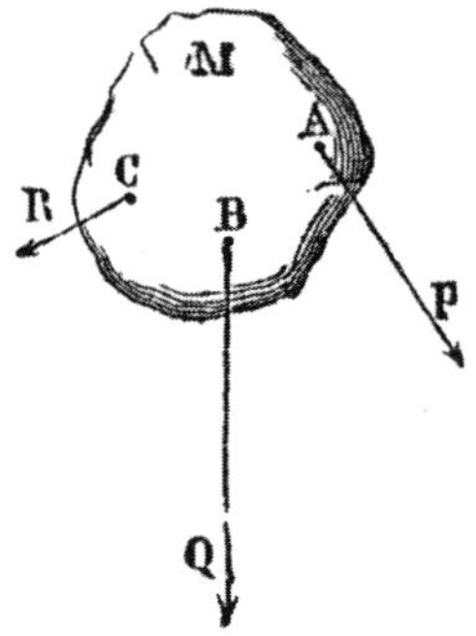

At the point A (fig. 5) a force is applied to the body, M, in the direction A P, while at B and C forces are applied in the directions B Q and C R respectively; we can thus represent the directions of the forces by straight lines, while arrow-heads on the lines show in which sense along these lines the forces tend.

§ 21. **Magnitude of a Force.**—We can also make the *length* of the line which indicates the direction of the force represent the magnitude of that force. For example, let it be arranged that a force of one pound shall be represented by a length of one inch, then a force of two pounds will be represented by a line two inches long, and so on. Thus the lines A P, B Q, C R in fig. 5 define in the most complete manner the forces which they represent. The line A P, for example, tells us that a force is applied at the point A of the body, that the magnitude of the force is proportional to A P, that it acts along the direction A P, and tends to draw A towards P.

§ 22. **Composition of Forces.** — Questions frequently arise in which it becomes necessary to consider the joint effect of two or more forces which act simultaneously upon a body. Let us suppose, for example, that *two forces are acting at the same point of a body and in the same direction.* Then it is plain that the joint effect of the forces is equal to the effect of a single force applied at the same point, provided the magnitude of the single force is equal to the sum of the magnitudes of the two separate forces. We have thus one force producing an effect which is equal to that produced by two forces.

If the two forces had not the same direction, then the operation is not quite so simple, but still it is generally possible to determine one force which is equivalent to the separate forces in the same plane. The process by which the single force is determined, which is equivalent to two or more given forces, is termed the *composition of forces.*

§ 23. **Parallel Forces.**—We shall first consider the case in which the two forces acting upon the body are *parallel.* For this purpose we shall employ the apparatus here described (fig. 6) :—

A uniform bar of wood A B is suspended at its middle point C in such a way that it is able to turn quite freely about the point of suspension. The front side of the bar has ten equidistant divisions (of, suppose, one inch each) on each side of the central point. Below the lines of division small rings are attached, for the purpose of suspending therefrom the weights, which will produce the forces applied to the bar. Each of these weights is furnished with a small hook both at the top and bottom, and weighs exactly one pound, or whatever other weight may be chosen as the standard. Since gravity acts in vertical lines, and since vertical lines are parallel, it is obvious that the forces applied to the bar by two weights suspended from it are parallel.

FIG. 6.

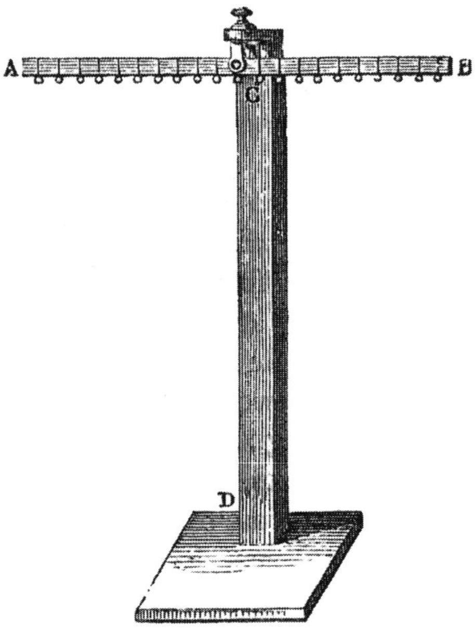

The first experiment may be made by suspending a weight of 1 lb. from each of two points equidistant from the centre of the bar. In the case shown in fig. 7 the two weights are each four inches distant from the centre. It is observed that the bar remains horizontal; there is in fact no reason why it should go down on one side more than on the other.

FIG. 7.

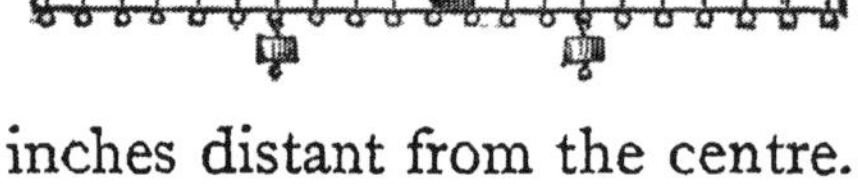

But if the two weights were removed and were suspended from the ring beneath the point of suspension, in the manner shown in fig. 8, it is obvious that the pressure of the bar upon its supports would be exactly the same as in the former case. We therefore see that the two forces of 1 lb. each which are applied in parallel directions to two points of the bar equidistant from the centre are precisely equivalent to a single force of 2 lbs. which acts at the centre of the bar.

FIG. 8.

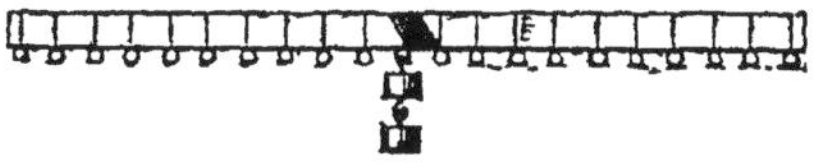

We thus learn in general that two equal and parallel forces applied to a rigid body and tending in the same direction may be replaced by a single force which is parallel to the separate forces, equal to their sum, and applied at the centre of the line joining the two points of application of the separate forces.

Suppose, however, that we suspend from the bar of fig. 6 eleven equal weights as shown in fig. 9, then, since the weights are uniformly distributed on each side of the point of suspension, it is manifest that the bar will remain at rest. If we remove the five weights on each side of the central weight, and suspend the whole eleven in the manner shown in fig. 10, the pressure of the bar on its supports must still remain the same as it was when the weights were ranged along the bar, and will be simply equal to the weight of the bar augmented by 11 lbs.

FIG. 9.

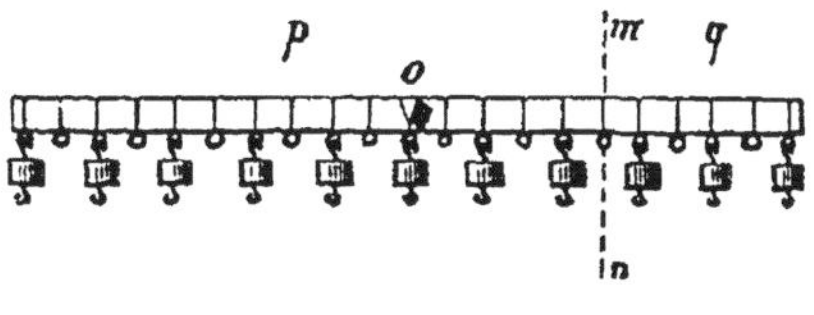

We may, however, deal with the weights in a different manner. Draw the line *m n* (fig. 9), which passes through the fifth division on the right-hand side of the bar. We thus have eight of the weights on the left of *m n*, and three of them on the right. But as the weights are symmetrically arranged in each of the two portions, we may proceed as follows. Observe that *p* is the centre of the left-hand portion, then taking off the eight weights two by two, we may hang them on at *p*, until the whole eight weights are suspended from *p*, and similarly the three weights on the right-hand side of the bar may be suspended from *q*, the arrangement then assumes the appearance shown in fig. 11, and it is found that the bar still remains horizontal.

Fig. 10.

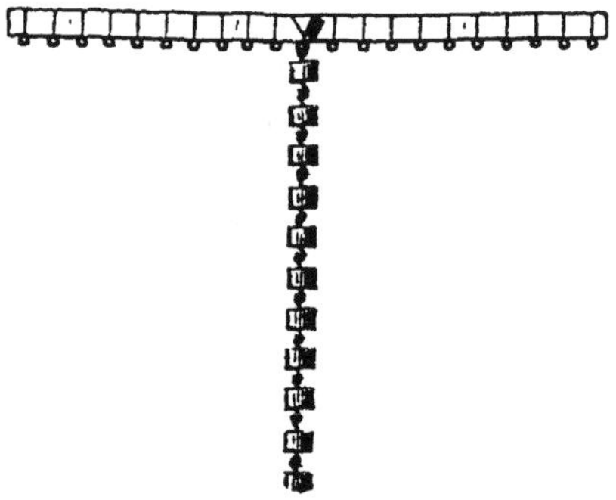

Comparing, then, the two cases shown in figs. 10 and 11, we see that the bar in each case remains horizontal, and that it supports the same total weight. We thus see that parallel forces of 8 lbs. and 3 lbs., at a distance of 11 inches, are equivalent to a single force of 11 lbs. applied at a point

Fig. 11.

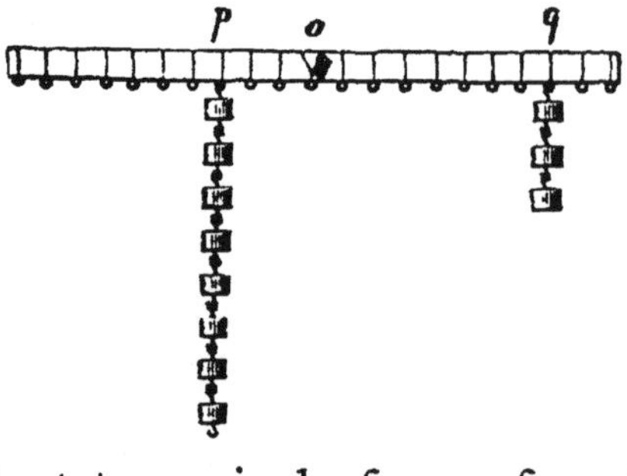

3 inches distant from the greater of the two component forces, and therefore 8 inches distant from the lesser.

This experiment is an illustration of the following general proposition.

Two parallel forces acting in the same direction on a rigid body are equivalent to a single force, which is parallel to them, equal to their sum, and is applied at a point which divides the line joining the two points of application of the single forces in the inverse proportion of those forces.

When a single force replaces two or more other forces, it is generally called their *resultant*, and the forces of which it is composed are called the *components.*

FIG. 12.

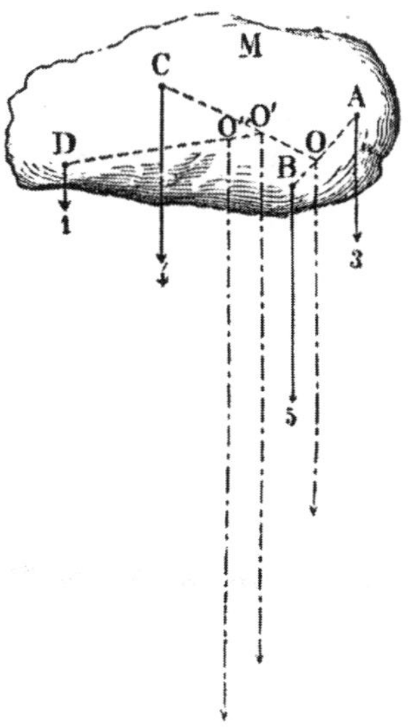

§ 24. **Example.**—Suppose a body M, fig. 12, be acted upon by four parallel forces, viz. a force of 3 lbs. applied at the point A : a force of 5 lbs. at the point B ; a force of 4 lbs. at the point C, and a force of 1 lb. at the point D. It is required to find the resultant. The two forces at A and B may be replaced by a single force of 8 lbs., which acts at the point O, where

O A : O B :: 5 : 3.

This new force of 8 lbs. at O may be combined with

the force at C, and the resultant will be a force of 11 lbs. applied at O′, where O O′ : C O′ :: 4 : 8. Finally the force at O′ may be similarly combined with the force at D, and the resultant of all the four forces is thus proved to be a single force of 13 lbs. applied at the point O″. We thus see how any number of parallel forces acting in the same direction can be compounded into a single force, and the resultant is always parallel to the component forces and equal to their sum.

§ 25. **The Lever.**—The very useful instrument known as the lever is, in one of its numerous forms,

FIG. 13.

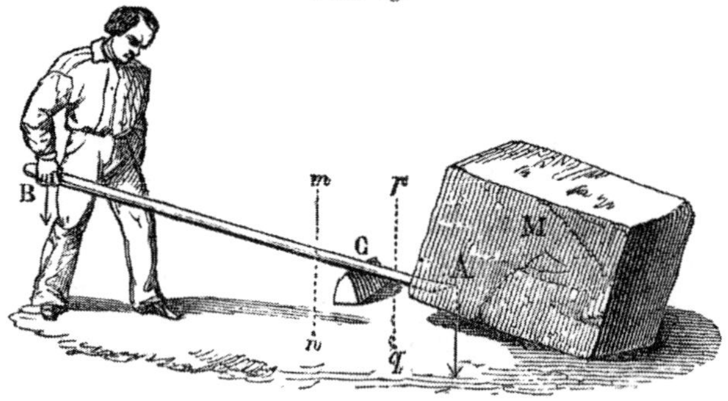

shown in fig. 13. This is a strong bar, A B, by the aid of which it is sought to raise a heavy body M, which bears upon the extremity A, while the power is exerted at the other extremity B. This bar presses at C on a support, about which it can turn when the power exerted at B is sufficiently great. When the heavy mass M is just kept sustained by the lever in the position shown in the figure, we may regard the lever as acted upon by two forces. The first of these is the pressure of the heavy mass M, which tends to force the extremity A of the lever downwards. The second force is the pressure exerted on the other end B by the hand of the workman, which is so regulated that it shall just sustain the load. These two forces may be regarded as parallel, and they can therefore be replaced by a single force, which will produce precisely the same effect upon the lever. Now this single force must pass through the point C, for if it lay either to the left of C, along the line *m n*, or to the right of C along *p q*, it would in either case tend to turn the lever round in the corresponding direction. From the fact that the lever is, as we have assumed, at rest, it follows that the resultant must pass through the point C. From this again it follows that the magnitudes of the forces at A and B must be inversely proportional to the distances of those points from C; thus if C B is ten times, a hundred times, a thousand times greater than C A, the pressure which must be exerted on B, will be one tenth, one hundredth, or one thousandth part of the load which presses on A.

§ 26. **Reaction.**—It is obvious that if two equal and opposite forces are applied to the same point of a body they will destroy each other, and will have no effect in altering the condition of the body as to rest or motion. Conversely, it is obvious that if a body be at rest, notwithstanding that one force is known to be acting upon it, we must infer that the effect of that force is neutralised by the effect of some other force. Apply this to the case of any object, a book for example, lying on the table. We know that the book is acted upon by the omnipresent force of gravity, yet the book is at rest. There must, therefore, be a force exerted by the table equal and opposite to the force of gravity. The force thus exerted is called *reaction.*

§ 27. **Equilibrium.**—When two or more forces are so related that they neutralise each other's effects on the objects to which they are applied, they are said to be in *equilibrium.* Thus the force of gravity and reaction of the table in the case of the book just considered, are said to be in equilibrium.

In the case of the lever of fig. 13, the forces also equilibrate, but it is necessary to observe that there is a third force acting on the lever besides the two forces already considered. This third force is the *reaction* of the support C. The pressures applied at A and B may be, as we have seen, compounded into a single force, equal to their sum, and applied at the point C. The reaction of the support O is a force precisely equal and opposite to this resultant. The reaction therefore neutralises the resultant, and the three forces are consequently in equilibrium.

Questions.

1. Explain how a force may be represented by a straight line?

2. If a straight line, 8 inches long, represent a force of 4 lbs., what is the magnitude of the force which is represented by a line 12 inches long?—*Ans.* : 6 lbs.

3. What is the magnitude of the resultant of two parallel forces which act in the same direction?

4. How do you find the position which the resultant occupies?

5. Two parallel forces of 9 lbs. and 7 lbs., at a distance of 32 inches, act upon a body. Determine the position and magnitude of the single force which will be equivalent to them?—*Ans.*: A force of 16 lbs., acting at a distance of 14 inches from the greater component.

6. Explain why the workman will have more power over the block in fig. 13 the nearer the points A and C are together.

7. What is meant by the force of reaction? What is the reaction of the table against a 2 lb. weight which rests upon it? *Ans.*: A force of 2 lbs. acting vertically upwards.

8. What is meant by forces being in equilibrium? When are two forces in equilibrium?

9. Point out the three forces which are in equilibrium in the case of the lever in fig. 13.

CHAPTER IV.

EQUILIBRIUM OF PARALLEL FORCES.

§ 28. **Dynamometer.**—For the purpose of measuring the intensity of a force, the ordinary spring balance, which is often used as a weighing-machine, is very convenient; a simple form of this apparatus is shown in fig. 14. It consists of a V-shaped piece of steel, and it is upon the elasticity of this steel that the action of the apparatus depends. A circular arc is firmly attached to the end of each branch of the spring. The free extremity of the one arc is furnished with a ring, while the other has a hook.

FIG. 14.

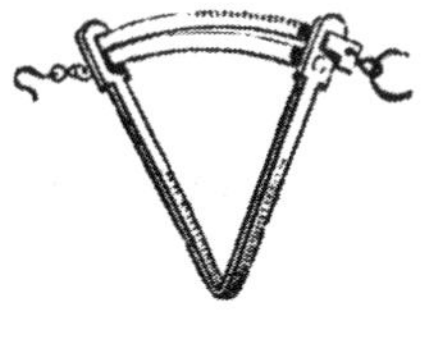

When a weight is suspended from the hook, the V-spring is gradually bent until the increasing elasticity of the spring is able to resist the weight, and it bends no further. By hanging different weights, it is found that the degree to which the spring is bent varies with the weight which is being supported. In this way the machine can be graduated by marks engraved on the circular arc, so that it can be used as a weighing machine.

If cords be attached to this apparatus, as shown in fig. 15, then the graduation of the instrument will show the tension of the cords or the amount of force which is being transmitted by the cord into which the instrument is inserted. When the spring balance is used in this way it is called a *dynamometer*.

FIG. 15.

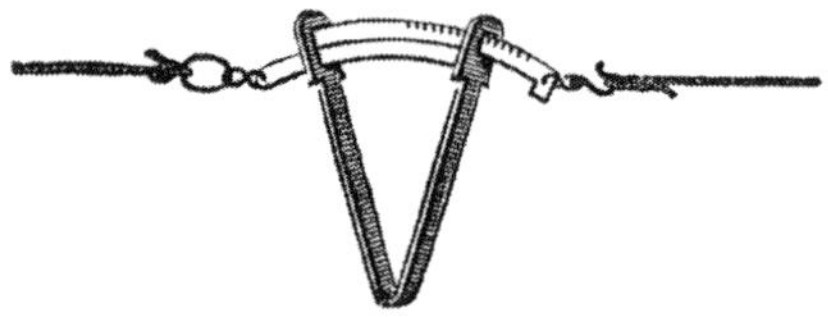

§ 29. **The Pulley.**—In the experiments about to be described, it will be necessary to make use of a *pulley* for the purpose of changing the direction of the force transmitted by a cord. The general construction of the pulley is shown in fig. 16. It consists of a wheel, mounted so as to turn very freely, with a groove in the margin to receive a cord. When the cord is pulled, the pulley revolves and the weight is drawn up.

FIG. 16.

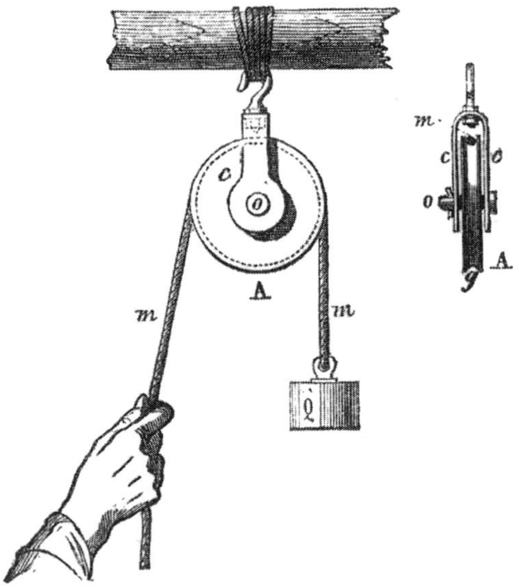

The main feature which makes the pulley of such

use in mechanics is the circumstance that the tension of a cord is almost unaltered when passing over a pulley. This may be illustrated by the apparatus shown in fig. 17. A string suspended from a fixed hook passes over a pulley at A, and carries a weight at P. In the two portions of the string on each side of the pulley dynamometers are inserted for the purpose of showing the tensions. It is then observed that the dynamometer *d'* shows that the lower part of the cord is subjected to a tension equal to the weight P, and that the tension of the upper portion of the cord, shown by the dynamometer *d*, is almost exactly equal to the weight P together with that of the lower dynamometer.

FIG. 17.

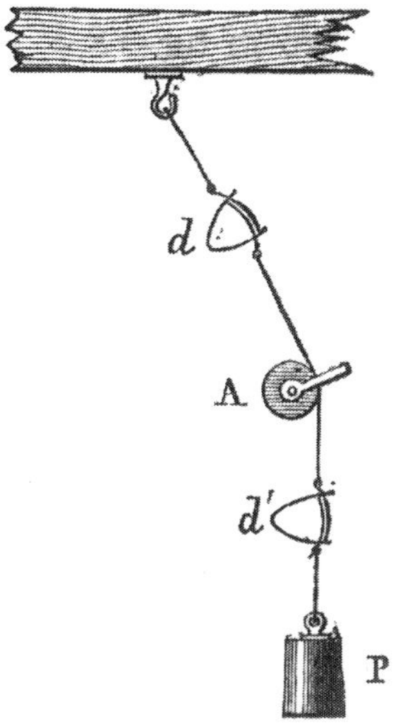

The two dynamometers would indicate that the tension due to the weight P is precisely the same in both parts of the cord, were it possible to make a perfect pulley (*i.e.* a pulley without friction), and also a cord which was perfectly flexible. These two conditions cannot be *perfectly* attained by any means at our disposal, but we can approximate to them so closely, that the difference of the indications of the two dynamometers may be insensible. In the experiments of which we are now about to speak, we may entirely overlook the minute differences arising from imperfect workmanship, and *we shall assume*

that the tension of a cord is absolutely unaltered when the direction of the cord is diverted by means of a pulley.

§ 30. **Experiment on Parallel Forces.**—We are now enabled to study experimentally the equilibrium of a body which is subjected to the action of three parallel forces. The apparatus employed for this purpose is shown in fig. 18. A body, H, I, K, is suspended from a dynamometer, which is hung from a fixed point. The weight of the body is of course shown by the indications of the dynamometer. Cords are attached to the body at the points H and I, which, after passing over two pulleys, support the weights P and Q, and for the sake of illustration, we shall suppose that P is 60 lbs. and that Q is 20 lbs. At the third point, K, another cord is attached, which, passing also over a pulley, supports a weight of 80 lbs.

When these arrangements have been made it is seen that the cord by which the body is suspended remains vertical as it was before, and the dynamometer shows that the tension of the cord is exactly the same as it was before the three forces were applied, and is in fact simply equal to the weight of the body. It is further noticed that the body assumes a definite position under the influence of the three forces, and that if it be displaced from that position it will, after a few oscillations, settle into it again. When in this definite position it is seen that the three cords A H, B I, and S K are all parallel; and regarding the weight of the body as simply counterbalanced by the tension of the cord by which the body is suspended from a fixed point, we have here a case of the equilibrium of three parallel forces.

In the first place it is to be noticed that the force of 80 lbs. on one side is equal to the sum of the two other forces of 20 lbs. and 60 lbs. The experiment has thus verified the law that whenever three parallel forces are in equilibrium one of these forces is equal to the sum of the other two. The experiment

FIG. 18.

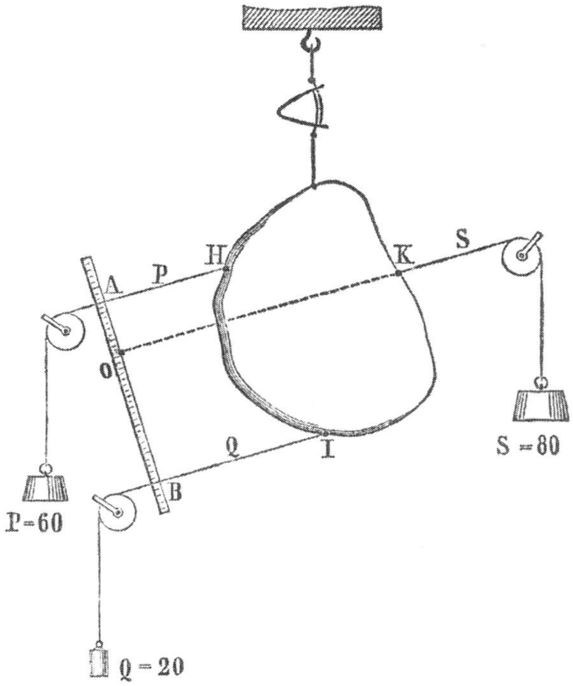

will, however, tell us more than this, for it can be made to show the relative positions which the three forces must occupy. For this purpose apply a divided scale in the manner shown at A B, then it is seen that the direction of the line S K cuts the scale A B in a point O, so that

$$OA : OB :: 1 : 3;$$

but the forces P and Q are such that

$$Q : P :: 1 : 3;$$

and therefore

$$OA : OB :: Q : P.$$

It is easy to see how this agrees with what we have already learned with respect to the composition of parallel forces. It appears from § 23 that two parallel forces of 60 and 20 must compound into a resultant of 80 acting at a point which divides the line joining the points of application of the components in the ratio of 20 to 60, or of 1 to 3. It follows that the forces along H A and I B must compound into a parallel resultant of 80 lbs. applied at the point O. This resultant is equilibrated by the precisely equal and opposite force directed along S K, and consequently the three forces equilibrate.

The principles we have been explaining are practically applied in a large number of useful contrivances, of which we shall now describe a few.

§ 31. **The Roman Balance.**—Fig. 19 represents a very useful form of weighing machine called the Roman balance, which depends for its action upon the laws of parallel forces. The machine consists of a strong rod which is suspended at the point C. The object to be weighed, P, is suspended from a hook attached to the rod at A, while the weight Q is moveable, and can be hung at any point on the arm C B. It will thus be seen that in this particular description of weighing machine the action depends not upon the magnitude of the counterpoising weight Q, for that magnitude is never changed, but on the position which it occupies on the bar C B when so placed that the bar is horizontal.

When the bar is laden in the way shown in the figure it is acted upon by three forces which equilibrate. Two of these forces are formed by the weights P and Q, while the third is communicated by the hand which holds the machine. These three forces are parallel to each other; and for equilibrium it is necessary that

FIG. 19.

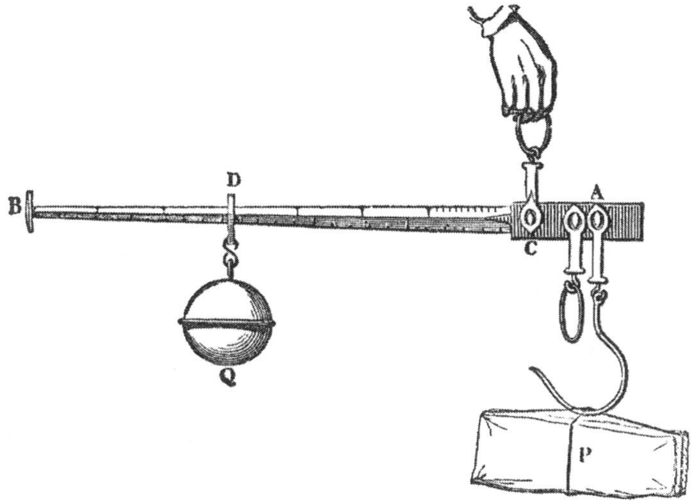

the two forces P and Q shall compound into a resultant equal and opposite to the third force. It is therefore necessary that the resultant of P and Q pass through C, whence we have the condition

$$P : Q :: CD : AC.$$

This may be written thus:

$$P : CD :: Q : AC.$$

Since Q and A C are both invariable when the machine

is once made, it follows that P must be always proportional to C D.

A scale constructed in the following manner is marked upon the bar C D. Take for P a weight of 1 lb., and ascertain the point at which Q must be hung so that the balance shall remain horizontal; this gives the unit of length according to which (setting out from C) the bar is to be divided.

In the case shown in fig. 19, Q is hung at the fourth division, and since P is proportional to C D, it follows that the weight of P must be 4 lbs.

FIG. 20.

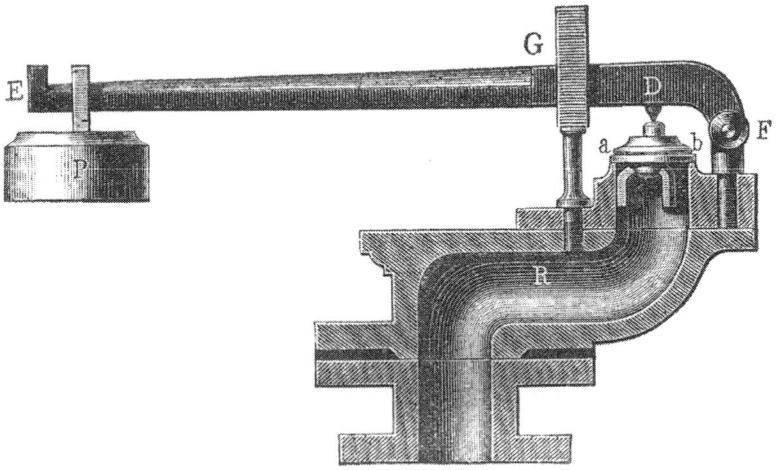

§ 32. **The Safety-valve.** — Among many other applications of the laws of parallel forces in the mechanical arts we may mention the *safety-valve*, with which the boilers of steam-engines are furnished. A section of a safety-valve is shown in fig. 20; and we shall now describe the principle upon which it acts.

The tube R, shown in the lower part of the figure, is in communication with the steam in the upper part of the boiler. This tube is accurately closed by the plug *a b*, which, however, merely acts like a stopper in a bottle, and would be immediately driven out by the pressure of the steam below. To keep the plug in its place it is pressed downwards at the point D by the strong arm F D E, which is hinged at F and loaded with a moveable weight, P. There is a guide at G for the purpose of keeping the arm in its proper place.

The pressure of the steam which tends to force the plug *a b* upwards is a vertical force, and this is really counteracted by two other forces which also act upon the arm F D E. One of these is, of course, the weight P; the other is the reaction of the hinge at F. The first is obviously a force directed vertically downwards, and it will also be seen that the same is true of the last; for the tendency of the end of the bar F is to fly upwards, and as this tendency is counteracted, the hinge must exert a force downwards. We are therefore led to regard the bar as acted upon by three parallel forces which make equilibrium.

It is evident that the downward reaction of the hinge must bear to the weight P a ratio equal to that of the segments in which the line joining their points of application is divided by the vertical line through D. According to the dimensions shown in the figure, this ratio is about 7 to 1, and therefore we infer that supposing the weight P to be 10 lbs. the reaction of the hinge must be 70 lbs. The resultant of these two forces will be equal to their sum, and therefore the total pressure on D, which is equal to that resultant, must be equal to 80 lbs.

It therefore follows that the steam will not be permitted to escape from the boiler unless it presses the plug upwards with a force of at least 80 lbs.

By shifting the weight P along the arm, the pressure brought to bear upon D can be changed. If, for example, P be moved in towards F, the pressure on D will be gradually decreased. It thus appears that the maximum pressure of the steam in the boiler can be controlled by the position which the weight P occupies along the arm of the safety-valve.

QUESTIONS.

1. A cord passes over a fixed pulley, and 1 lb. is suspended from each of its ends. What is the tension of the cord? *Ans.*: 1 lb.

2. Describe the Roman balance. If the distance D C, fig. 19, be 3 inches, when a weight of 9 lbs. is hanging from P, what will be the weight at P when the distance of Q from C is 9 inches? *Ans.*: 27 lbs.

3. In a Roman balance the distance A D is 28 inches, and it is found that Q is 3 lbs. and P is 11 lbs., what must be the distance A C? *Ans.*: 6 inches.

CHAPTER V.

THE PARALLELOGRAM OF FORCE.

§ 33. **Equilibrium of Three Forces.**—A particle O (fig. 21), is acted upon by three forces, *P*, *Q*, and *R*, in the directions O P, O Q, O R. These forces are all equal in magnitude, and they are so arranged that the angles P O Q, Q O R, and R O P, are all equal. Now it is manifest that under these circumstances, the point O will remain at rest, and the three forces will equilibrate. For if the forces do not equilibrate, the point O must begin to move, but since the forces and the angles are all equal, why should the point move in one direction rather than in another? If the point be not at rest, it must either move along the direction of one of the forces, or into the angle between two forces, but why should it move in the direction of one force rather than in that of another, or why should it move into one of the angles rather than another? The only inference is, that the point O does not move, and that the three forces are in equilibrium.

FIG. 21.

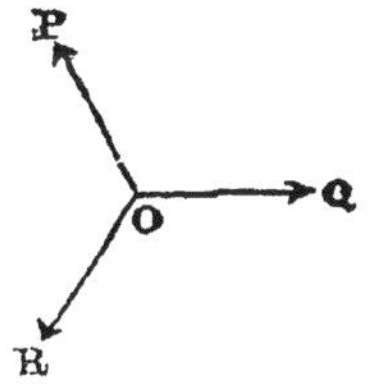

It therefore appears that each force is neutralised by the joint effect of the two remaining forces. Thus, for example, the two forces, *P* and *R*, taken together, completely destroy *Q*.

FIG. 22.

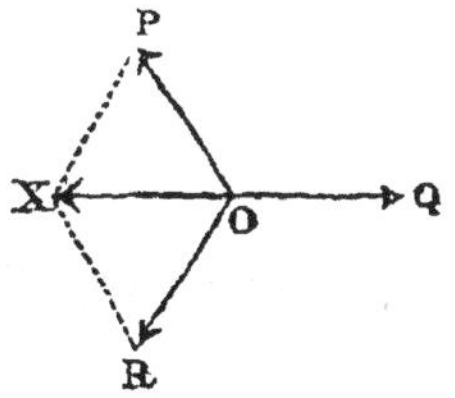

Produce the line Q O to X (fig. 22), so that O X

is equal to O Q. Now it is evident that if a force X equal to Q acted at O along O X, it would equilibrate Q along O Q, because they are equal and opposite; but we have seen that P and R equilibrate with Q, hence we must admit that the combined effect of the two forces P and R is precisely the same as that of the single force X.

We thus see that the two forces, P and R, may be replaced by the single force X, and we therefore say that X is the *resultant* of P and R.

It may be asked, how it is that one force can produce the same effect as two forces, each of them as large as itself. The apparent paradox is explained by observing that P and R, though not exactly opposite, are still to a certain extent pulling against each other. They therefore partly destroy each other, and what is left of them is equivalent to X, both in magnitude and in direction.

Join P X and R X, then the simplest geometrical consideration will show that the figure O R X P is a parallelogram. For since the angle P O Q is two-thirds of two right angles, it follows that the angle X O P is one-third of two right angles, and that therefore the triangle X O P, which is obviously isosceles, is also equilateral. Similar considerations apply to the triangle X O R, and hence the four sides of the figure O R X P are all equal, and consequently it must be a parallelogram.

We here see that the resultant of the two forces P and R is represented both in magnitude and direction, by the diagonal of the parallelogram, of which O P and O R are two conterminous sides.

§ 34. **Resultant of Two Equal Forces.**—Let us now take the case of any two equal forces, *A* and *B*, which act upon the point O (fig. 23) in the directions O A and O B. It will be observed that this case is more general than the case just considered, inasmuch as the angle between the two forces is now supposed to have any value whatever.

FIG 23.

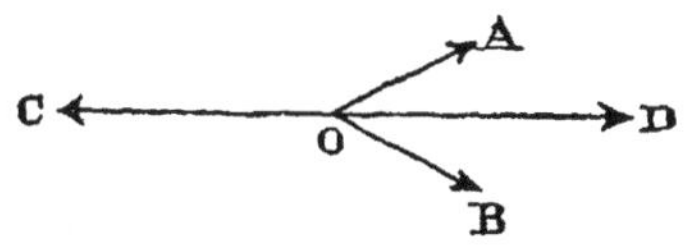

It is clear that the joint effect of the two forces, *A* and *B*, must be to move the particle O, and it is further obvious, from the symmetry of the case, that as these two forces are equal, the direction in which they tend to move O, must lie in the bisector O D of the angle A O B. There is, in fact, not the slightest reason why O should move to one side of O D rather than to the other.

If O is to be held in equilibrium by three forces, of which *A* and *B* are two, then it is necessary that the third force shall act in a direction opposite to O D, and have a certain magnitude which may be represented by O C. We thus have the two forces, *A* and *B*, equilibrated by a third force *C* acting along O C. It is, however, plain that *C* would be equilibrated by an equal and opposite force *D* acting along O D. Hence the joint effect of the two forces, *A* and *B*, is equivalent to a single force *D*, acting in the direction of the line bisecting them.

§ 35. **Resultant of any Two Forces.**—Passing to the most general case, suppose that any two forces, *P* and *Q*, act upon a particle at the point O (fig. 24). In general the effect of these forces would be to make the particle move off in a direction which we denote by O T. Then it is plain that a force *R*, in the same line as O T, could be applied so as to prevent the particle from moving towards T and retain it in the original position O.

FIG. 24.

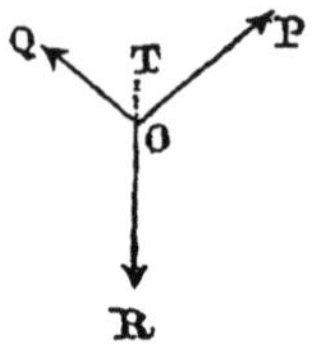

Under these circumstances the three forces, *P*, *Q*, *R*, are in equilibrium. But the force *R* would of course be destroyed by a force equal and opposite to it, and therefore it follows that a force equal and opposite to *R* would be equivalent to the joint effect of two forces, *P* and *Q*, and form what we have called their *resultant*.

§ 36. **The Parallelogram of Force.**—The results we have already obtained may be summed up in the general law now to be enunciated (fig. 25).

The resultant R of two forces, P and Q, acting concurrently on the same point, is obtained by marking off on the directions of the two forces, lengths which represent those forces according to a certain scale, then forming the parallelogram, of which these lengths are two conterminous sides, and drawing the diagonal through their intersection. The direction of this diagonal is the direction of the required resultant, and the length of this diagonal represents the magnitude of the resultant. Example: Let the two forces be $P = 60$ lbs. and $Q = 20$ lbs. respectively, and let their directions be inclined at an angle

FIG 25.

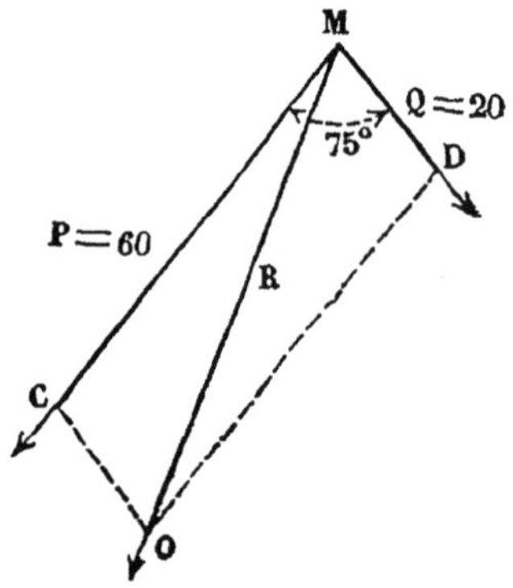

of 75°. By constructing the diagram shown in fig. 25 and measuring the diagonal, it is found that the magnitude of the resultant is 68 lbs., while measurement with a protractor shows that the angle C M O, at which the resultant is inclined to the larger force, is equal to 16° 30′.

§ 37. **Experimental Illustration.**—The doctrine of the parallelogram of force can be easily verified experimentally. Let A and B (fig. 26) be two small pulleys fixed to a wall. These pulleys should be very nicely made, so that there shall be as little friction as possible, and so that consequently the tension of the flexible silken cords passing over them shall be as little altered as possible. Let P, Q, R, be three weights attached to three silken cords, the other ends of which are all knotted together at *c*. The cord attached to the weight P is to be passed over the pulley A, while the cord attached to Q passes over the pulley B, the cord attached to R hangs vertically downwards from C. The knot at *c* is therefore acted upon by three forces, directed along *c* A, *c* B, *c* R, and respectively equal to the weights of P, Q, R.

FIG. 26.

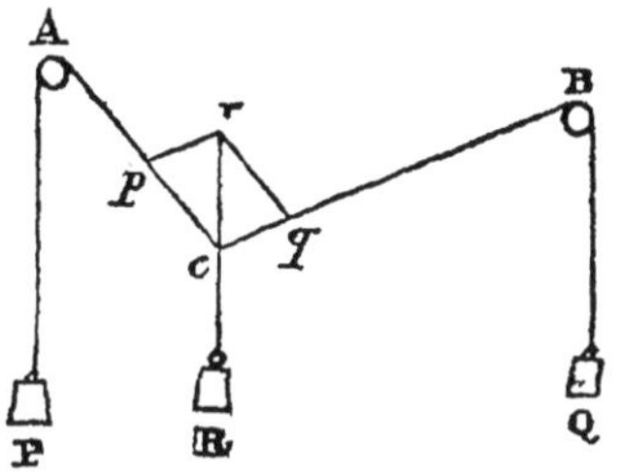

When these arrangements are made, it is found that the cords assume a definite position, and that if the knot be displaced it settles again after a few oscillations into its original place. We then mark on the wall lengths *c p* and *c q*, each containing as many inches as there are ounces in the weights P and Q respectively, and complete the parallelogram *c p q r*. It is found that the diagonal *c r* is vertical, and that the number of inches in the length *c r* is equal to the number of ounces in the weight at R. This verifies the parallelogram of force, for the diagonal *c r* represents a force which is capable of supporting R. It therefore has the same effect as the forces represented by *c p* and *c q*, and consequently it is their resultant.

§ 38. **Resolution of Forces.**—By the aid of the principle of the parallelogram of force, we are enabled to solve the converse problem of resolving one force into two other forces, of which the original force is the resultant.

Let O S (fig. 27) represent a force applied at the point O. Draw through O *any* two lines O P and O Q, and then through S draw lines S Q, S P, parallel to O P and O Q. It appears by the parallelogram of force that the forces represented by O P and O Q would compound into a resultant represented by O S; hence conversely the given force represented by O S may be decomposed into the two separate forces represented by O P and O Q. The two forces into which a given force is resolved are often termed the *components* of the given force.

FIG. 27.

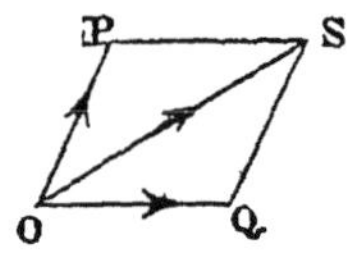

§ 39. **Illustrations.**—A simple experiment (fig. 28) will illustrate the method of resolving a force into two components. A cord is attached to a point A, fastened to the wall, while the other extremity B of the cord is tied to the hook of the dynamometer. At a point C in the cord A B, another cord, C D, is knotted, which at its lower extremity carries the weight D. The point C is then acted upon by three forces, which make equilibrium. The first of these forces is due to the weight D, and acts vertically downwards along the line C D. The second force acts along the cord C A, and really arises from the reaction of the wall at the point A. The third force is produced by the elasticity of the dynamometer, and acts along the line C B.

Produce C D vertically upwards to E. Mark off

FIG. 28.

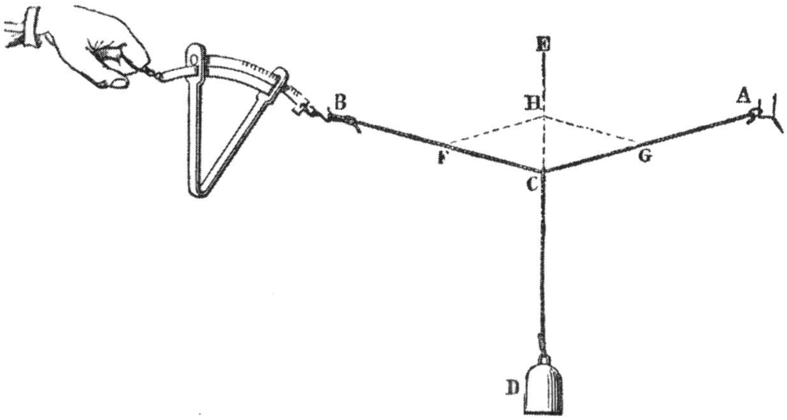

on C E a length C H, which contains as many inches as there are ounces in the weight D. Through H draw H F parallel to A C, and H G parallel to B C. Then the lines C F and C G will denote on the same scale the forces along C A and C B. The dynamometer affords a ready means of testing this result, so far as the force along B C is concerned. If we measure the number of inches in the length C F, then the principle of the parallelogram of force asserts that the force in that line is the same number of ounces. That this is the case appears at once from the indications of the dynamometer.

§ 40. **The Jib and Tie.**—A very useful application of the principles now under consideration is presented in the jib and tie, shown in fig. 29. A weight, P, is suspended from a point C, which is supported by the jib A C and the tie B C. The jib consists of a stiff piece of wood, which is capable of turning round a joint at A. The tie is a piece of cord attached at the points B and C; a dynamometer may be inserted in it, for the purpose of showing the tension to which the cord is subjected. The upright support A B is fixed firmly.

Fig. 29.

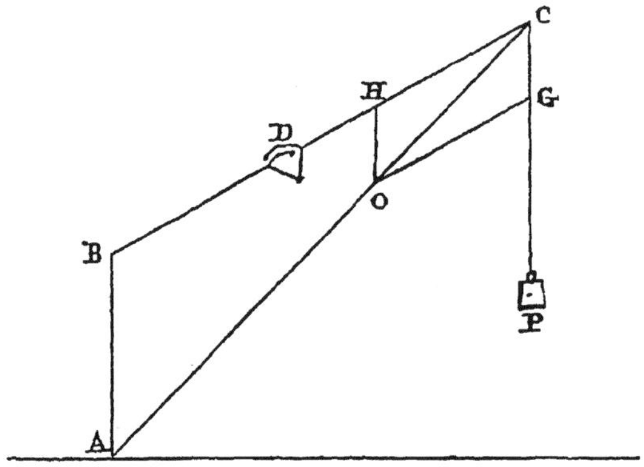

The point C is then in equilibrium under the action of three forces. The first of these is the weight at P, which acts vertically downwards; the next force is really the reaction of the hinge at A, which acts along the jib A C, and tends to push the point C outwards. The third force is the tension of the tie C B. A moment's reflection will show that the forces along the tie and the jib are really in the directions we have just indicated. For suppose the tie were to be cut, what would happen? The point C would, of course, fall away to the right under the influence of P; that is, C would tend to go farther away from B than it is at present. Or suppose the jib were to lose its stiffness, it is evident that the apparatus would collapse, and C would come nearer to A than it was before. It is therefore manifest that the effect of the jib is to push C outwards.

Take a length C G, which contains as many inches as there are units in the weight P; draw through G a line G O, parallel to C H, and then draw through O a line O H parallel to C G. The parallelogram C G O H is thus formed, and it appears, from the parallelogram of force, that the force exerted by the jib is represented by O C, while the tension of the tie is represented by C H. The presence of the dynamometer in the tie enables us to verify that the number of inches in the length C H is equal to the number of units in the strain shown by the dynamometer.

§ 41. **Sailing of a Ship.**—A somewhat more complicated illustration of the resolution of forces is afforded by the action of the wind on the sails of a ship. In the case of a ship sailing with the wind directly astern the matter is simple enough. Let X Y be the direction of the ship (fig. 30), and A B the direction in which the sail is set. If the wind blows in the direction P X, it presses upon the surface of the sails, and produces a force, which, acting at O, urges the ship forwards.

FIG. 30.

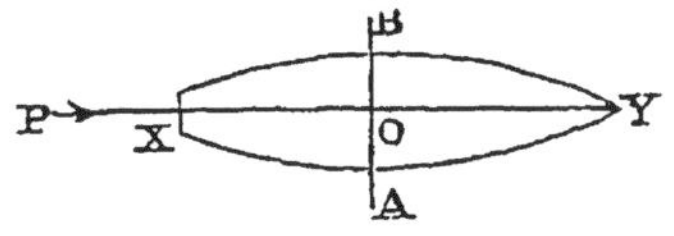

Next suppose that the wind is blowing from the north, while the course of the vessel is east. It is well known that the sails can be so set that the ship shall be driven on its course, notwithstanding the fact that the direction of the wind is perpendicular to the direction of the motion of the ship.

Let P O (fig. 31) be the direction of the wind, while X Y is the direction of the ship's course. It is clear that it would not answer in this case to set the sails in a direction which was perpendicular to the direction of the wind, for the only effect of the wind would then be to blow the vessel in the direction O Q, which is not wanted. But if the sails are set in the direction A B, then the ship can be steered so as to move in the required direction X Y.

FIG. 31.

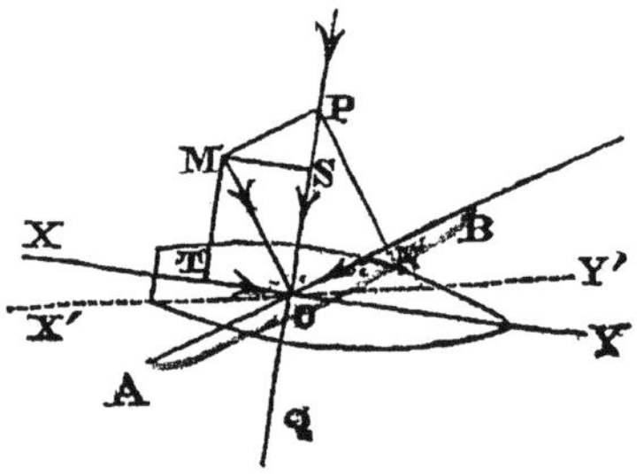

For let the line P O represent the magnitude of the force exerted by the wind. Draw through O a line O M perpendicular to A B, and through P draw P N and P M respectively parallel to O M and O B. Then P N O M is a parallelogram, and therefore the force denoted by P O may be replaced by the two forces N O and M O. We may therefore consider the original wind to be replaced by two winds, one blowing along A B, and exerting a force denoted by N O, while the other blows perpendicular to A B, with a force denoted by M O. It is obvious that the first of these winds can have no effect whatever, as it merely runs along the sail without pressing upon it. We may therefore discard this component entirely, and consider only the wind M O, which blowing perpendicularly upon the sail, produces its full effect.

The immediate effect of the force M O would be to force the ship in the direction M O, but this tendency is modified by the resistance which the water opposes to an object of the shape of the ship. To make this clear, it is necessary to resolve O M into two other components, which is done by drawing lines M S and M T parallel respectively to O X and O P. We may then replace the force M O by the two forces S O and T O. The action of the original wind has therefore assumed the form of the two forces S O and T O, and we must consider the effect of these forces separately.

The force denoted by S O tends to push the ship in a direction perpendicular to that in which the ship is intended to go. This force is to be resisted as far as possible, and with this object the shape of a ship is so designed that the water will oppose very great resistance to the motion of the ship sideways. In spite, however, of the shape of the ship, the force S O would produce some effect in driving the vessel to leeward, if it were not practically counteracted by steering the vessel, along a line X′ Y′, which is slightly inclined to the actual course of the vessel, X Y.

The really effective portion of the wind is represented by the line T O. This tends to force the vessel along in the direction in which it is intended to go. The shape of the vessel is so designed that the resistance with which it is opposed by the water shall be as small as possible in this direction.

QUESTIONS.

1. Two forces of 3 lbs. and 4 lbs. act on a point in directions at right angles to each other. Find the resultant.

> *Ans.* : When the parallelogram is drawn, it is seen to be a rectangle, and therefore the square of the diagonal is equal to the sum of the squares of the two adjacent sides 3 and 4. Since $3^2+4^2=5^2$ it is evident that the resultant is 5 lbs.

2. Show that when the directions of two forces intersect at right angles they and their resultant are proportional to the sides of a right angled triangle.

3. Prove that when three forces equilibrate at a point the sum of any two of these forces must be greater than the third force.

4. If a force of 10 lbs. be resolved into two equal components which are at right angles, what must be the magnitude of the components? *Ans.* : $10 \div \sqrt{2}$ lbs.

5. If in fig. 29 the distances A C, C B, B A be respectively 21 feet, 16 feet, 9 feet, find the tension of the tie and the thrust on the jib, when P is a load of 1,800 lbs.

> *Ans.* : It is evident that the triangle O C H is similar to the triangle A C B, and consequently O H : C H :: A B : C B. Hence the tension of the tie bears to P the same ratio which B C bears to A B, and thus the answer is easily seen to be 3,200 lbs. Also the force on the jib bears to P the same ratio as A C : A B, and hence the thrust on the jib is 4,200 lbs.

CHAPTER VI.

GRAVITATION.

§ 42. **Gravitation towards the Earth's centre.**— It is a matter of experience all over the earth, that a heavy body tends to fall to the ground.

Fig. 32 represents a section of the earth which passes through its centre O. Then the earth being sufficiently near a sphere for our purpose, we may regard the section P Q R S, as a circle where P Q R S are the intersections of the lines O A, O B, O C, O D, with the surface of the earth.

FIG. 32.

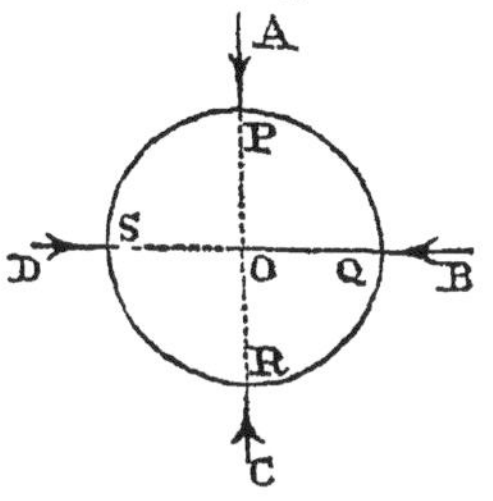

If a stone be dropped from a point A, above the surface of the earth, it will fall to the ground at P, but it is to be observed that if the stone were dropped from B or C or D, it would descend to the earth at Q or R or S respectively. In each case the stone will try to reach the earth by the shortest path. From A the stone appears to fall downwards, from C it appears to move upwards, from B it moves to the left, and from D it moves to the right.

The four movements, however, have one feature in common, they all tend towards the centre of the earth, and hence we deduce the important result which is thus stated :

A body dropped from a point above the surface of the earth always falls in a straight line, which is directed to the centre of the earth.

§ 43. **Attraction.**—The familiar instance of a magnet and a small piece of iron will suffice to illustrate what is meant by the word *attraction.* In virtue of certain properties possessed by the iron and the magnet they are drawn together. The magnet draws the iron and the iron draws the magnet. This particular kind of attraction is, however, of a very special character; the magnet appears to have no appreciable influence on a piece of wood or paper for example, and has, indeed, no considerable influence on any known substance except iron and one or two metals closely allied to iron by their chemical characters.

There is, however, a kind of attraction which is called the *attraction of gravitation.* This is of a different character from the attraction of a magnet for iron. According to the law of gravitation, every body attracts every other body, *whatever be the materials of which it is composed.* The intensity of this attraction is, however, so small that with one conspicuous exception, we can only become aware of its existence by refined and elaborate inquiries. That any two objects—for example, two books lying on the table—do actually attract each other, there can be no doubt whatever; but the intensity of this force is so small that the attractive force cannot overcome the friction of the table, and consequently we do not find that the books are drawn together. The larger and more massive the two bodies, the greater is the intensity of the force of gravitation between them. Though this force is intrinsically so small, yet the mass of the earth is so large that the attraction which it exercises upon an object near its surface is quite appreciable.

It is this attraction of the earth which produces what we call *weight*, and it is the attraction of the earth which causes bodies to fall to the ground.

Let O be the earth's centre (supposed to be a sphere), and let A be the position of a stone. As the earth attracts the stone, it must evidently draw it along the line A P. There is, in fact, no reason why the stone should move to one side of the line more than to another. We thus see that attraction would always tend to draw objects in a direction pointing towards the centre of the earth. The observed facts are therefore explained by the supposition that the earth possesses the power of attraction.

§ 44. **The Centre of Gravity.**—In studying the effect of gravity upon a body, it will be convenient to commence with a simple illustrative experiment which can be easily tried.

Out of a piece of cardboard or tin plate, a figure of *any shape*, A, B, C, D, E (fig. 33) is to be cut. In this plate a few holes are to be punched, quite at random, as, for example, near A, B, C, D, E. In the wall is a nail, by which the plate can be suspended by passing the nail through any of the holes, for example, A. From the nail is suspended in front of the plate a cord A H, which is kept in the vertical direction by the plummet attached to it at H.

FIG. 33.

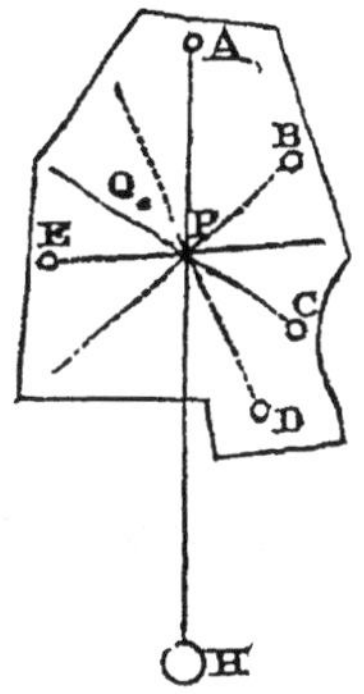

As the plate is not supported in any other way, it hangs quite freely from the nail, and if I displace the plate it will, after a few oscillations, settle down again in the position which it occupied at the first. In order to mark this position, it is desirable to draw a line on the plate in the direction A P, indicated by the plummet line, which is hanging in front. If the plate be blackened, this can be neatly done by chalking the plummet line, and then giving it a flip against the plate.

When the line has been drawn, remove the plate and hang it up again from some other hole, for example, B. Again the plate settles into a definite position, and again the line and plummet is to be hung on, and a second line drawn in the direction of the plummet. The two lines drawn on the plate intersect at a point P.

When the plate is hung up on a third hole, for example, C, and a third plummet line is drawn, a very remarkable result is perceived. It is found that the line drawn on the third occasion passes through the intersection of the two former lines, that is to say, all the three lines pass through the point P.

Repeating the operation with other holes, D, E, &c., it is found that all the lines drawn in the way we have described possess the remarkable property of passing through one definite point of the plate.

It is therefore manifest that the point P is characterised in a very remarkable way. It is only stating the results at which we have arrived in a somewhat different manner to assert that from whatever point the plate be suspended, the point P must be vertically beneath the point of suspension when the plate is hanging at rest.

If we punch a hole through the plate at P, and then hang the plate on the nail by means of that hole, we see another remarkable result. *The plate will remain suspended indifferently in any position whatever.* The significance of this observation will be much enhanced if we make a hole at a point Q near to P. When the plate is suspended from Q, it will only be at rest in one position, that is, when P is vertically beneath Q.

It must surely be regarded as a matter worthy of close attention that any plate, of any figure, regular or irregular, should have one specific point in it which enjoys the unique properties which the experiments show are possessed by the point P. This point has received a name, it is called the *centre of gravity*.

In the illustration we have just given, we have spoken merely of a thin plate, because the experiments are more easily conducted in a body of this nature than in one of entirely irregular form. It must not, however, be supposed that a plate is the only kind of body which possesses a point having the properties we have described. No matter what be the shape of a body it possesses a centre of gravity.

FIG. 34.

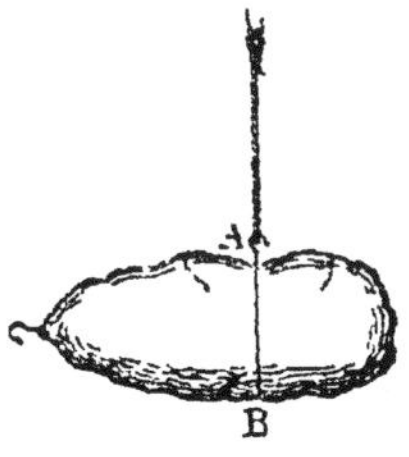

§ 45. **Centre of Gravity of a Body of any Shape.**—A body of any shape or materials whatever, A B C D (fig. 34), is suspended by a cord from a point A, and hangs in a certain position when at rest. We may suppose that a vertical hole, A B, is drilled through the body in the direction of the cord by which the body is suspended. If we now suspend the body by another point on its surface, C, the body will come to rest in the position which is represented in fig. 35. Let us now conceive that a second hole is drilled through the body in the direction of the line C D. It will be found that these two straight holes intersect in the interior of the body. In fact, if we thrust a knitting-needle through one of the holes, and then attempt to thrust another knitting-needle through the other hole, we shall find that the way is stopped in the interior of the body by the first knitting-needle.

FIG. 35.

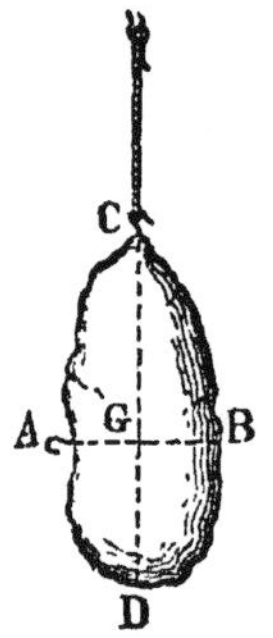

If the body be now suspended from any other point on its surface, and if a similar hole be made through the point, and in the direction of the string by which the body is suspended, it will be found that this hole also passes through the intersection of the two former holes.

From each and every point of suspension the same result is obtained, and thus we are led to the conclusion that in a body of any shape or form whatever there is one point which possesses the remarkable property thus stated :

When a body suspended by a cord from a fixed point is at rest, there is one special point of the body which is always vertically beneath the point of suspension, whatever may be the point of the body to which the cord is attached.

This point is the *centre of gravity.*

§ 46. **Position of the Centre of Gravity.**—In the case of a homogeneous body (i.e. a body composed of uniform materials throughout) of regular shape, the centre of gravity is determined from the most simple considerations of symmetry.

In the case of a sphere, it is obvious that the centre of gravity must lie at its centre, for there is no other point symmetrically related to the figure.

In the case of the right circular cylinder (fig. 36), the centre of gravity must be the central point of the axis of the cylinder, for it must obviously lie *somewhere* on the axis of the cylinder, and there is no reason why it should lie more to one end of the cylinder than the other. It must, therefore, be at the centre.

FIG. 36.

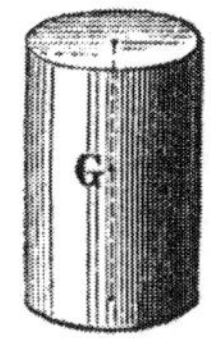

It is, perhaps, slightly more difficult to see that in the case of the oblique cylinder (fig. 37) the centre of gravity must also lie at the central point of the axis, G. It will, however, be seen that a point in any other position would not be symmetrically placed with respect to the material of the body.

FIG. 37.

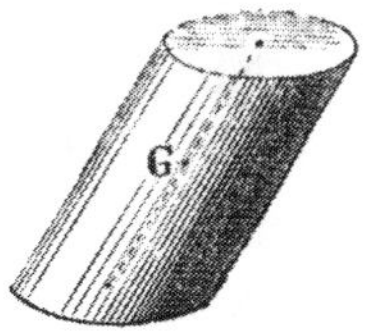

In a similar way the centre of gravity of the parallelopiped (fig. 38) is situated at the centre of its volume. This is found by joining two opposite corners of the figure, and thus making a diagonal; joining another pair of opposite corners, we have a second diagonal; and the intersection of these two lines gives the point G, which is the centre of gravity of the mass.

FIG. 38.

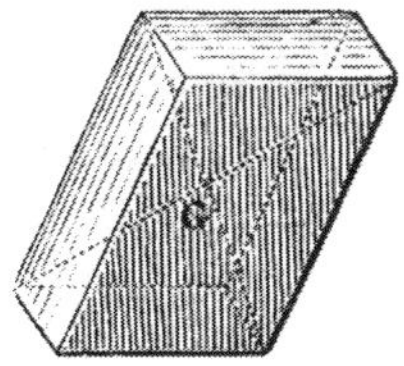

It will sometimes happen that the centre of gravity

of a body may not lie in the substance of the body at all. A very simple case of this is presented in a ring (fig. 39). It is obvious from symmetry that the centre of gravity cannot be elsewhere than at the centre, G, which is the centre of the figure.

FIG. 39.

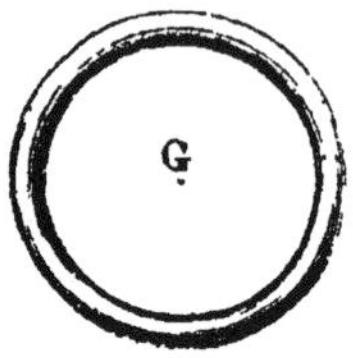

§ 47. **Property of the Centre of Gravity.**—A body of any description may be considered to be composed of an innumerable multitude of small particles of matter. Each of these particles is acted upon by the attraction of the earth. Each particle is therefore urged towards the earth by a certain force which tends towards the earth's centre.

Considering that the centre of the earth is nearly 4,000 miles distant, the directions of these forces may, for all practical purposes, be regarded as parallel. Even if the two particles were a mile distant, the inclination of the directions of the two forces would be scarcely a minute. We may therefore regard the forces as absolutely parallel whenever a body of comparatively small dimensions is under consideration.

We have seen that two parallel forces may be compounded into a single force which is parallel to the two components; therefore the forces on two particles of the body may be replaced by a single force. This force may be similarly compounded with the force on a third particle of the body, this resultant with the force on a fourth particle, and so on, indefinitely, until all the particles have been included.

We thus come to the important conclusion that the effect of gravity upon a body of any shape is to produce *one force*, which acts vertically downwards.

It remains to be shown that *this force acts through the centre of gravity of the body.* Take a body of any shape (fig. 40), and suspend it by the cord O A; then, when the body is at rest, the centre of gravity, G, must lie in the direction O A, vertically beneath the point of suspension. If the direction of that one force which constitutes the effect of gravity does not pass through the point G, let us suppose that it lies along the line P Q. The body is then acted upon by two forces. These are, first, the weight due to the action of gravity, and secondly, the reaction of the cord which acts along the line O A. As the body is at rest, these two forces must be equal and opposite. It is therefore plain that the action of gravity cannot lie along the line P Q, but must lie in the same straight line as O A, and must, therefore, pass through the point G.

FIG. 40.

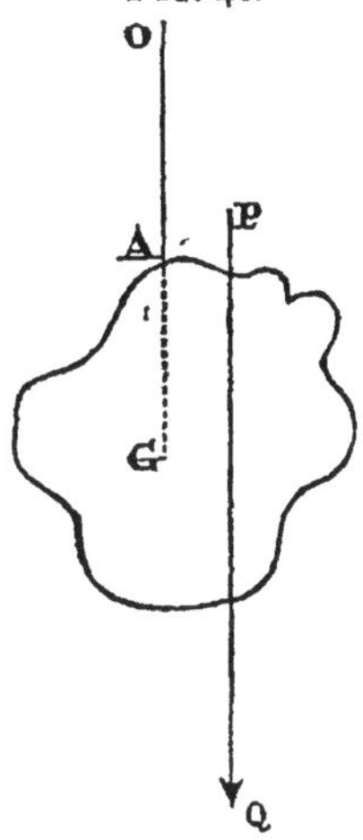

We are therefore led to the very important conclusion that *the effect of the attraction of the earth upon a body is to produce a single force, which passes through the centre of gravity of the body, and acts vertically downwards.*

§ 48. **Stability of a Body resting upon a Horizontal Plane.**—The properties of the centre of gravity will enable us to render a satisfactory explanation of a great number of observed facts, of which we shall here mention a few.

The accompanying figures 41 and 42 each represent a cylinder placed upon a table; there is, however, an important difference between the two cases, for whereas the cylinder in fig. 41 will remain steadily standing, that in fig. 42 will tumble over. What is the cause of the different behaviour of the cylinder in the two cases?

FIG. 41.

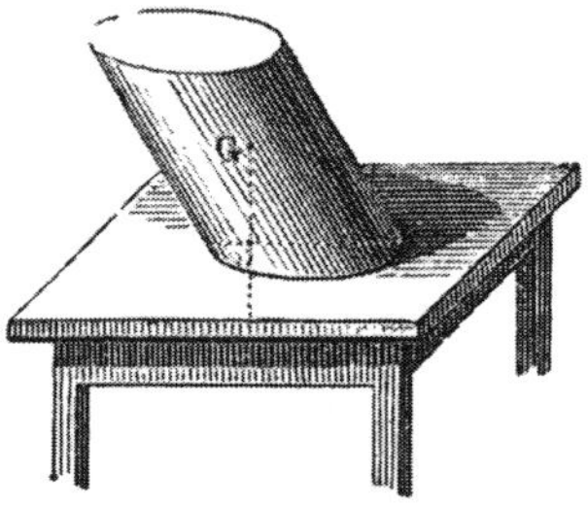

We have already seen that in a cylinder of this form the centre of gravity is at the centre of its axis. Now the entire effect of gravity is to produce a force acting vertically downwards through the centre of gravity. Therefore in each case the entire effect of gravity is to produce on the cylinder a vertical force acting downwards through the point G. The vertical lines through G possess different features in the two cases. In fig. 41 the vertical through G passes through the base of support, i.e. the part of the cylinder which is in contact with the plane. In fig. 42, on the other hand, the vertical through G falls outside the base of support. In each case the cylinder is acted upon by the force of gravity and the reaction of the table. If these two forces equilibrate, then the cylinder

FIG. 42.

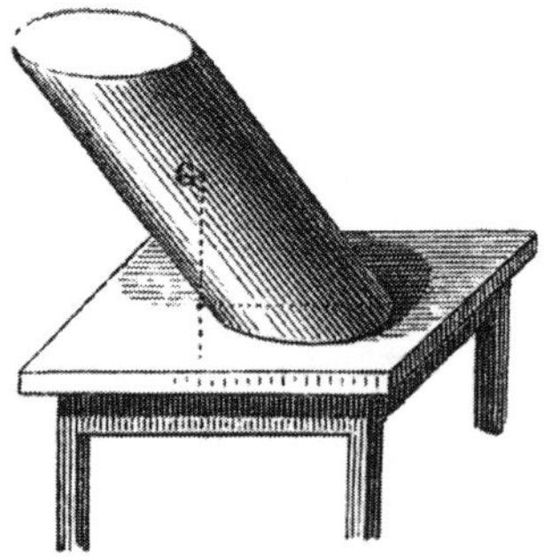

will remain at rest. It is, however, plain that the reaction of the table can only act at some point or points where it is in contact with the cylinder. It is therefore impossible that the reaction of the table can be exactly opposite to the force through G in the case of fig. 42, and consequently equilibrium is impossible, and the body tumbles over.

But in the case of fig. 41 matters are different. It is obvious that the force of gravity would tend in this case to keep the body steadily on its base rather than to tumble it over. We are therefore led to the following general statement:

Whenever the centre of gravity of a body is vertically over the base by which it is supported, the body will remain in equilibrium.

§ 49. **Stable and Unstable Equilibrium.**—Although a body reposing on a horizontal plane is at rest when the vertical through the centre of gravity intersects the base of support, yet this circumstance will not always guarantee the permanency of the equilibrium. Take, for the sake of illustration, an oval body, A B C D (fig. 43).

FIG. 43.

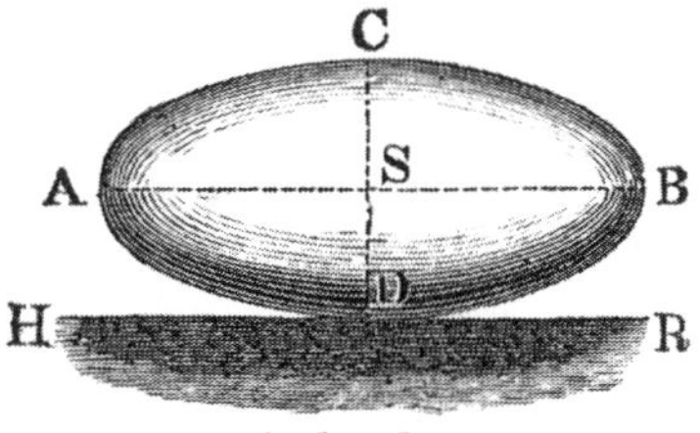

The centre of gravity is at the point S, and the body is now at rest with its shorter diameter, C D, vertical. If the body receive a slight displacement under such circumstances, it is obvious that after a few oscillations it will return to its original position. In this case the equilibrium is said to be *stable.*

It is, however, conceivable that the same body

could be balanced with its longer diameter, A B, vertical, as shown in fig. 44. In this case, no doubt, the centre of gravity, S, is vertically above the point of support, A; and so far the condition of equilibrium is fulfilled. But there is a wide difference between the character of the equilibrium in this case and in the other case. We now observe that if the body receive the slightest displacement it will not recover itself as it did before, but it will tumble over entirely. We therefore call equilibrium of this kind *unstable.*

FIG. 44.

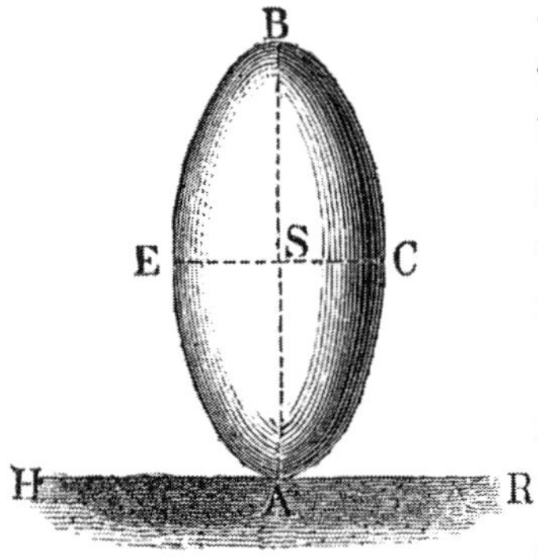

Intermediate between the cases of unstable and stable equilibrium we have what is called *neutral* equilibrium. This may be illustrated by a sphere placed upon a horizontal plane (fig. 45). In this case, when the sphere receives a slight derangement from the position in which it was previously lying at rest, it makes no effort either to return to its original position, as in the case of stable equilibrium, or to depart farther therefrom, as in the case of unstable equilibrium.

FIG. 45.

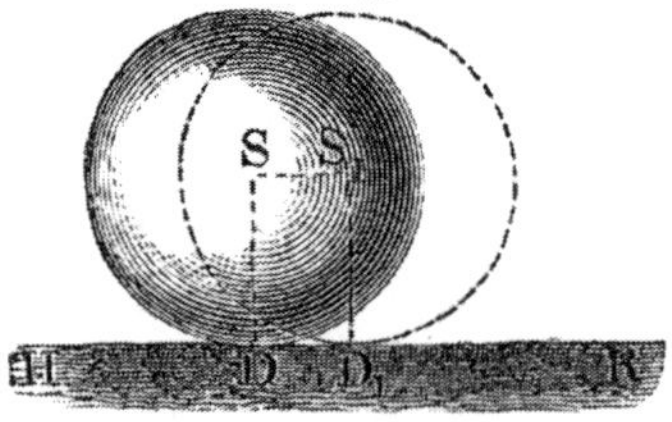

It will be noticed that the centre of gravity of sphere S moves into the position S_1, when the sphere is deranged. It is now again vertically over the new point of support D_1, and therefore the new position is equally a position of equilibrium with the original position. We may notice that here the height of the centre of gravity above the horizontal plane remains unaltered. This will give the clue to the discrimination between equilibrium which is stable and that which is unstable. In fig. 43 a slight displacement of the figure obviously causes the point S to rise; on the other hand, a slight displacement of the oval in fig. 44 lowers the point S. We are therefore led to the following statement:

When the centre of gravity is at its lowest point the equilibrium is stable, and when the centre of gravity is at its highest point the equilibrium is unstable.

The different descriptions of equilibrium may also be illustrated by fig. 46. The rectangle in the first

FIG. 46.

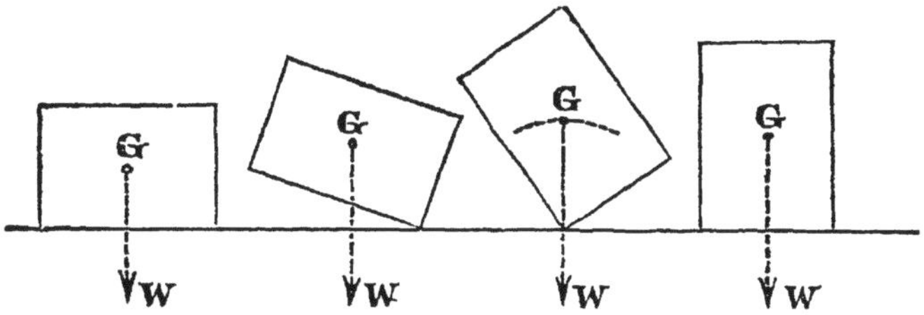

figure is obviously in stable equilibrium, for if it be raised up in the manner shown in the second figure it will return to its original position when released. If, however, the rectangle be tilted sufficiently high, it may attain the position shown in the third figure, where, the centre of gravity having reached its highest point, the rectangle is in unstable equilibrium; after passing this position the rectangle may assume the second position of stable equilibrium shown in the fourth figure.

Fig. 47.

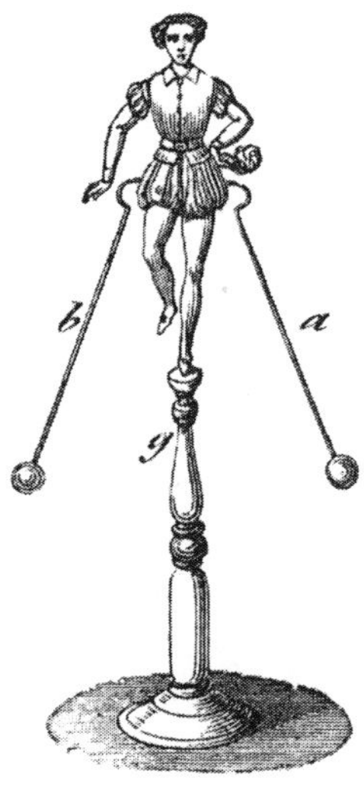

In the toy represented in fig. 47 the centre of gravity of the man's figure is above the point on which his foot is supported, and therefore his equilibrium in this position would be unstable. By means of the rods *a*, *b*, two heavy leaden balls are attached to his figure; they have the effect of bringing the centre of gravity so low down as to be beneath the point of support. In this case the equilibrium is stable, and the man's figure may be displaced very considerably from the vertical, and still it will return thereto.

Questions.

1. Find the centre of gravity of the triangular plate A B C, fig. 48.

Fig. 48.

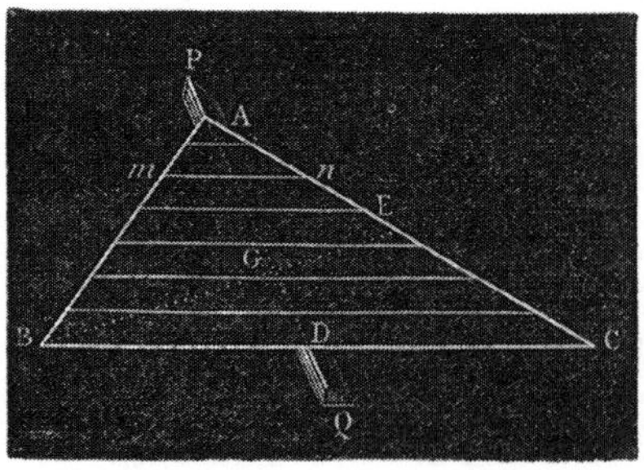

Bisect the base B C of the triangle in the point D, then it can easily be proved that a line A D, drawn from the vertex of the triangle to the point of bisection of the base would bisect every line parallel to the base, such as *m n*. Let a large number of lines be drawn parallel to the base, and suppose that the triangle is placed so that A D lies along the edge of a prism P Q. If the strips into which the triangle is divided by these lines are so excessively narrow that we may regard them merely as slender rods, it is evident that each of the strips would balance itself separately on the edge of the prism, because the middle point of the strip is over the edge of the prism. As each of the strips would balance by itself, it is plain that the whole triangle which is formed of these strips will balance, and, therefore, the centre of gravity of the triangle must lie on the line A D (§ 48).

In the same way it can be shown that the centre of gravity must lie in the line B E, which is drawn from the vertex B to bisect the side A C. The centre of gravity must, therefore, lie at the point of intersection of these two bisectors, i.e. at the point G.

2. A triangular plate is to be carried by three men, each of whom holds one of the corners. If the plate weighs 45 lbs., how much of the weight will each man have to bear?

Ans. The weight of the plate is a vertical force of 45 lbs., which may be regarded as acting at the centre of gravity G, fig. 49. Join B G and produce it to D. It can easily be proved that the point G, which is the intersection of two bisectors of the sides drawn from the opposite angles, is the point of trisection of the line B D. It follows from this that the force of 45 lbs., acting at G, is equivalent to two parallel forces of 15 lbs. and 30 lbs., acting at B and D respectively. But since D is the central point of A C, a force of 30 lbs. acting at D is equivalent to two parallel forces, each of 15 lbs., acting at A and C. The entire weight of the triangle will, therefore, be equally divided among the three men.

FIG. 49.

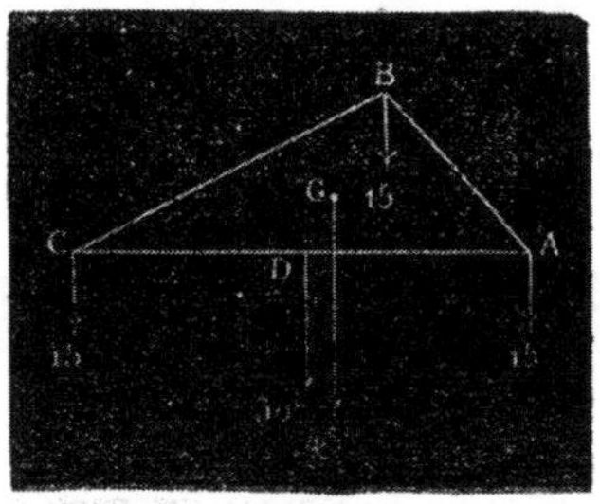

3. Two particles, weighing 3 and 5 oz. respectively, are at-

tached to the ends of a straight rod 8 inches long, weighing 2 oz. Find the centre of gravity of the system.

Ans. The weights of the two particles may be combined into a single force of 8 oz., which acts at a distance of one inch from the centre of the rod. This force may be compounded with that due to the weight of the rod, which acts at its centre. The total action of gravity upon the rod and the particles, therefore, constitutes a force acting at a distance of four-fifths of an inch from the centre. Since this force must pass through the centre of gravity, it follows that the centre of gravity must be four-fifths of an inch from the centre of the rod.

4. What condition must be fulfilled in order that a body may rest in equilibrium upon a smooth plane surface?

5. Give examples of bodies in stable, unstable, and neutral equilibrium. If a body be in stable equilibrium, how is its centre of gravity affected by a small displacement of the body?

6. Prove that the centre of gravity of three equal bodies placed at the corners of a triangle coincides with the centre of gravity of the area of the triangle.

7. When a heavy body is suspended from a point about which it can turn freely, show that it will rest with its centre of gravity in the vertical line passing through the point of suspension.

8. Show that a body cannot be in stable equilibrium on a horizontal plane if it rests on less than three points of support.

CHAPTER VII.

ILLUSTRATIONS OF THE EQUILIBRIUM OF FORCES.

§ 50. **Distribution of the Load upon a Beam between the two Points of Support.**—A beam A B (fig. 50), supported at its extremities on two walls of the same height, is laden in the point O with a certain load R. It is required to find the pressures upon the points of support.

FIG. 50.

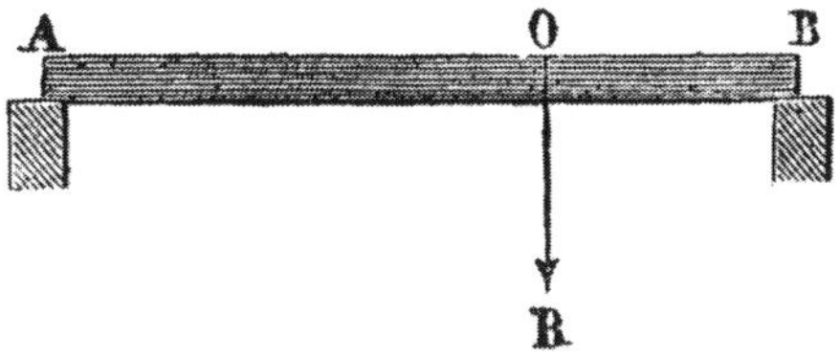

It is sufficiently plain that if the weight were applied at the centre of the beam, each of the two supports would have to bear half the load. If the load were placed exactly over one of the supports, then that support would have to bear the entire load, and the other support would have no pressure upon it beyond that due to the beam. When the load occupies any other position, as for example that shown in the figure, then each support will have to sustain a certain share of the load.

We may conceive that the supports are removed, provided that we replace them by two suitable forces acting vertically upwards. The effect of each support is therefore to press the beam upwards with a certain force. The pressure which each end of the beam exerts against the support is equal and opposite to the reaction of that support against the beam. We can therefore determine the pressures on the supports, by the laws which regulate the equilibrium of a body acted upon by three parallel forces.

For the sake of illustration, let us suppose that the beam A B is 12 feet long, and that a load of 1,200 lbs. is applied at the point O, which is 4 feet distant from the end B. In order for the beam to be in equilibrium, the intensities of the two forces at the extremities of the beam must be in the inverse ratio of the segments into which the load divides the beam. The pressure at B must therefore be double the pressure at A, and as the sum of these two pressures is 1,200 lbs., it is plain that the pressure at B must be 800 lbs., and the pressure at A 400 lbs.

It is of course needful to add to each of the pressures here found half the weight of the beam itself, to get the total pressure on each support.

§ 51. **Funicular Polygon.**—A flexible cord, K M H, (fig. 51), is fastened at two points, K and H, to nails in a wall, and at any point M, a third cord is knotted, from which a weight R is suspended. It is required to find the tensions on the parts of the cords M H and M K.

FIG. 51.

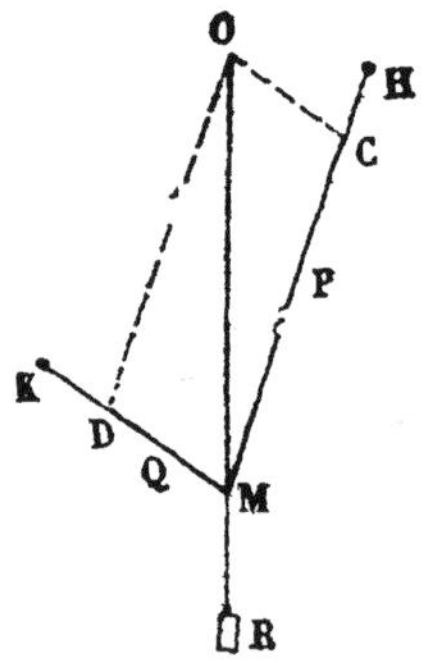

This problem may be solved by geometrical construction, in the following manner. Produce R M upwards to O, and take as many inches in the length M O as there are ounces in the weight R. Through O draw O C parallel to M K, and O D parallel to M H.

By the parallelogram of force, it appears that tensions in the cords M K and M H of intensities M D and M C, will compound into a resultant M O, which is equal and opposite to the weight R.

Hence the number of inches in M D is equal to the number of ounces in the tension of M K, and the number of inches in M C is equal to the number of ounces in the tension of M H.

The figure assumed by a flexible cord from which weights are suspended, is often termed the *funicular polygon.* The present case is an elementary illustration.

It may be noticed that the tension of the cord, M H, exceeds that of the cord M K, and that the former is more nearly vertical than the latter.

The greater the angle K M H the greater are the tensions of the two cords relatively to the weight R.

FIG. 52.

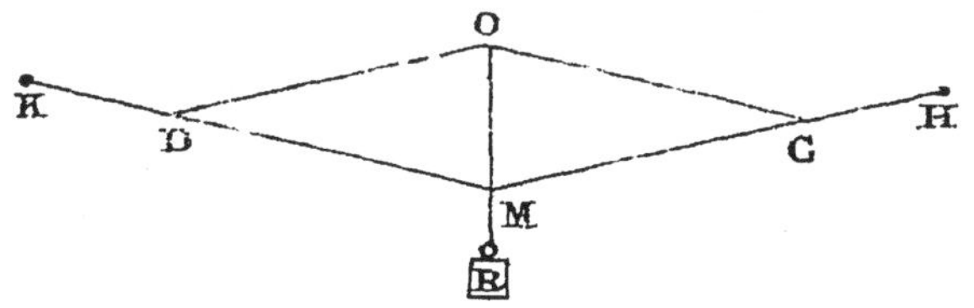

This is illustrated in fig. 52, where the ratios of M C and M D to M O are obviously much larger than in the previous case. It appears from this, that if a cord stretched pretty tightly between the two points K and H be deflected, even by a very small weight applied transversely, a very great tension will be produced in the cord.

This is illustrated by the case of a telegraph wire suspended from pole to pole, in the way with which every one is familiar. The wire is never in a straight line, it is always curved downwards in the centre. Even though the wire is very tightly stretched when it is put up, it is impossible to make it perfectly horizontal. In fact the wire, strong as it is, would be broken before this could be accomplished. This is explained by recollecting that the wire has weight. This weight constitutes a force acting vertically downwards, and therefore produces a tension on the wire, which increases in proportion as the wire approaches the horizontal. The wire then could not remain perfectly horizontal, because in this case its tension would be indefinitely great, and fracture must ensue.

The student may ask how it comes to pass that if the weight of a body can be represented by a single force applied at its centre of gravity, the form of a telegraph wire should really be a curve, and not rather an angle, like that produced in the cord of fig. 52 by a weight hung at its centre. The reply to this is that the weight of a body acts as though it were concentrated in a single point only, *when that body is rigid*, and as the wire is flexible, instead of rigid, its weight cannot be adequately represented by supposing it to constitute a single force at the centre of the wire.

§ 52. **The Bent Lever.**—In discussing the lever, we assumed that it was a straight rod, and that it was submitted to the action of forces whose directions were parallel. We now consider the case of the bent lever A C B (fig. 53), which, turning around the point C, is acted upon by forces at A and B, which are perpendicular to the corresponding arms A C and B C. We may imagine the lever A C to be prolonged to B′, so that C B′ is equal to C B; if then a force be applied at B′ in a direction perpendicular to B′ C, and equal to the force applied at B, it will be obvious that these two forces have equal effects in turning the

FIG. 53.

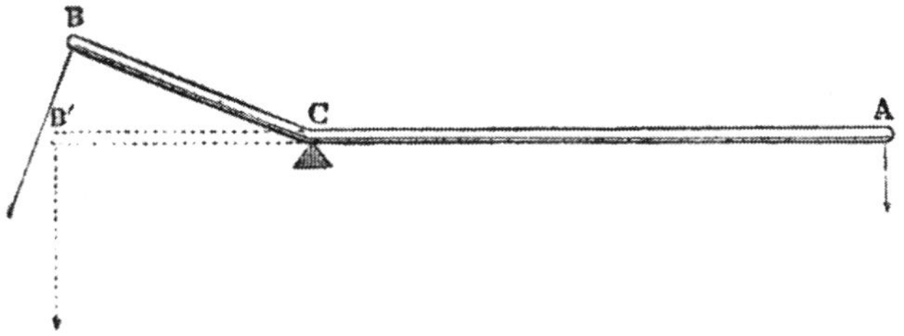

lever round the point of support, C. But the straight lever A C B′ will be in equilibrium, provided the forces at A and B′ are inversely proportional to the arms of the lever on which they act, whence it follows that the bent lever A C B will be in equilibrium, provided that the forces at A and at B are in the inverse proportion of the lengths A C and B C.

It sometimes happens that the forces applied at the extremities of a bent lever are not perpendicular

FIG. 54.

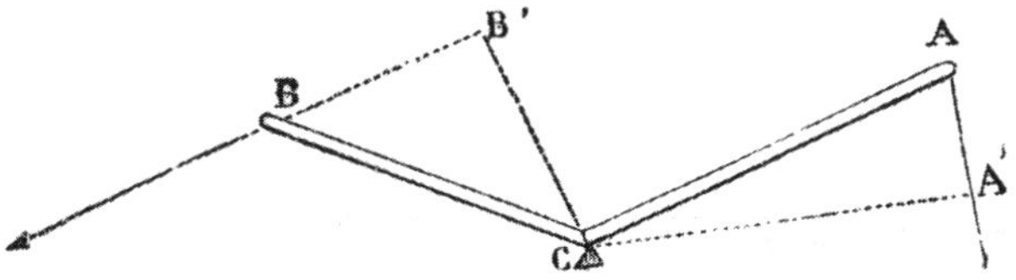

to the arms of the lever. In a case like this (fig. 54) we draw lines C B′ and C A′ from the point C perpendicular to the directions of the forces. Then the given forces acting at A and B may be replaced by equal forces acting at the points A′ and B′ on the bent lever A′ C B′; and for equilibrium it is only necessary that the intensities of these forces be inversely proportional to the lengths of the perpendiculars C B′ and C A′.

We may state the principle of the lever in the following more general way, which will include the case just considered:

Two forces acting upon a lever are in equilibrium, when the intensities of the forces are inversely proportional to the perpendiculars let fall from the point about which the lever turns on the directions of the forces.

This principle may be illustrated by the following experiment. A circular disc (fig. 55) is mounted so that it can turn freely about a fixed axis through the centre C. At any two points A, B, of the disc cords are attached which, after passing over the fixed pulleys L, M, support weights P, Q. The disc is observed to assume a certain definite position under the influence of the two forces which are applied to it at A and B.

FIG. 55.

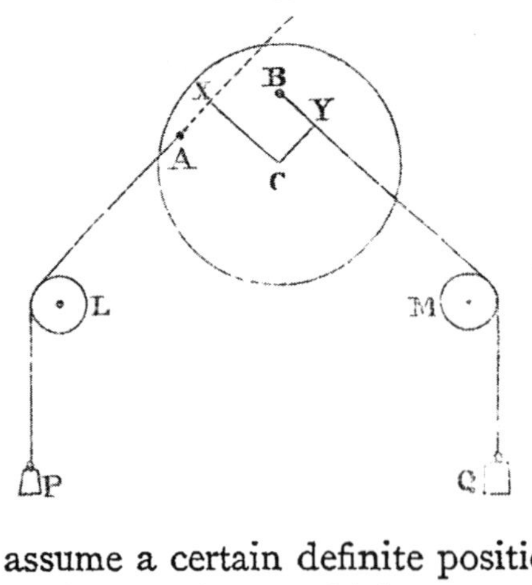

From C let fall perpendiculars, C X, C Y, upon the directions of the cords (produced if necessary). If, then, these perpendiculars be measured, it is invariably found that their lengths are inversely proportional to the weights P, Q. In this case the disc performs the part of the bent lever; and it is convenient to adopt this form, because the weight of the disc itself is entirely supported at C, and therefore does not complicate the question.

It is often convenient to express the relation between the forces and the perpendiculars by the equation

$$P \times \text{CX} = Q \times \text{CY}.$$

§ 53. **The Balance.**—The balance is an apparatus for weighing bodies; that is, for determining the number of grains, ounces, or pounds that the bodies contain. The most essential part of the instrument is a lever, often called the beam (fig. 56), which is supported at its centre, and from the ends of which two pans are hung, in one of which the object to be weighed is placed, while the other contains the weights with which it is to be compared.

It is important that this beam be freely moveable about its point of support, and that the point of support be exactly at the centre of its length. With this view the beam is furnished with a steel edge, which is fixed transversely at its centre, and projects out on each side. These projecting edges rest upon two small horizontal planes of agate or other hard material, one in front of the lever and the other behind it. When the beam is oscillating, it turns about these edges.

At the extremities of the beam two other steel edges are fixed; these edges are parallel to the central one, which is half-way between them; and it is essential that the three edges be in the same horizontal plane. The edges at the ends of the beam support hooks, from which the pans are suspended by wires or chains.

In order that a balance shall discharge its functions properly, it is necessary that it shall be both *just* and *sensitive*.

FIG. 56.

A balance is said to be *just* if its beam remains horizontal whenever the pans are either empty or contain weights which are known to be equal. This requires that the central steel edge should be placed exactly half-way between the two steel edges at the ends of the beam, and that further the two arms of the beam on each side of the central edge, with their appended hooks, chains, and pans, should be perfectly symmetrical. In this case it is obvious that, when the beam is horizontal, the centre of gravity of the beam, chains, and pans, will lie in the vertical plane, passing through the central edge, and consequently the balance will be in equilibrium.

If equal weights are placed in the two pans they will not disturb the equilibrium, because the two arms of the lever are equal, and therefore the two weights compound into a single vertical force which acts at the centre of the beam, and is counteracted by the reaction of the agate plate on which the edge rests.

If the pans of a just balance be laden with two weights which are not equal, the beam will no longer remain horizontal, but will be inclined to the horizontal by a certain angle. The size of this angle will of course depend upon the difference between the loads; but it will also depend to a certain extent upon the construction of the balance itself. For example, in a balance of coarse construction, when the pans are each laden with a weight of 1 lb. the beam is of course horizontal, and the addition of a grain to one side will hardly render the inclination of the beam to the horizon perceptible. But in a balance of a superior kind the addition of a grain, or even of a small portion of a grain, will produce a very appreciable inclination of the beam. In this case we say that the balance has greater *sensibility* than in the former case.

The sensibility of a balance is therefore measured by the angle of inclination of the beam which can be produced by a given excess of weight in one pan over the weight in the other pan.

The sensibility depends to a certain extent upon the friction of the axis against its support. It is therefore desirable that the steel edge should be delicately and carefully made of very hard steel, and that the plate on which it rests should be perfectly smooth. But the sensibility of a balance also depends upon the degree of accuracy with which the two following conditions have been fulfilled :

1st. That the edges of suspension of the beam and the pans lie in the saam plane.

2nd. That when the beam is placed horizontally, its centre of gravity shall be vertically beneath the edge of suspension, and very close thereto.

The advantages of these arrangements are shown as follows :

Let A B (fig. 57) be the beam of a balance, O its centre of suspension, and G its centre of gravity. Then when an excess of weight, p, is placed upon the pan which is suspended from A, the beam will move into the position A′ B′.

FIG. 57.

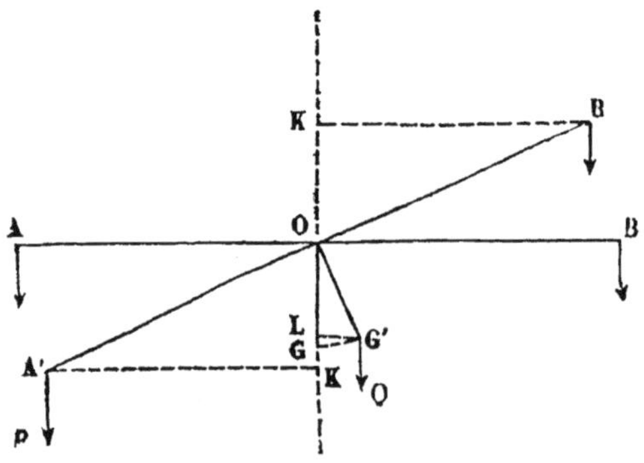

Since the points A′ O B′ are in a straight line, of which O is the centre, it follows that two equal weights suspended from A′ and B′ will compound into a vertical force passing through O; this force will be counteracted by the reaction of the supports, and may be left out of consideration. It therefore appears that the first of the conditions we have mentioned provides that the balance shall have the same degree of sensibility whatever be the weights placed upon the pans, provided that their difference is the same.

The equilibrium of the beam, after it has assumed the inclined position A′ B′, will depend upon the weight Q of the beam itself acting at the point G′ and upon p, acting at A′, and equal to the difference between the weights in the two pans. These two forces must therefore equilibrate when applied to the bent lever A′ O G′. Letting fall perpendiculars A′ K and G′ L from A′ and G′ on the vertical line O K we deduce the equation (§ 52)

$$p \times A'K = Q \times G'L.$$

It is obvious from the figure that the nearer G is to O the smaller will the arm of the lever G′ L become. It therefore appears that A′ K must be smaller according as G is nearer O, whence we see that for a given weight p the angle A O A′ will increase rapidly according as the centre of gravity of the beam is nearer the axis of suspension. The advantage of the second of the two conditions for increasing sensibility is therefore obvious.

If the point G be too far from O, then the balance is not sufficiently sensitive; if G were actually at O, then it would be too sensitive, for the minutest difference between the weights in the pans would cause the beam to become vertical. If G were above O, then the balance would be unstable, and the minutest difference between the weights in the pans would upset the beam.

A pointer fixed to the beam moves in front of a graduated arc. This shows the position which the beam occupies, or the angle through which it has been displaced.

To test the justness of a balance it is sufficient to place a certain weight in one pan and balance it by putting sand or any similar substance in the other pan. Then interchange the weights, and if the beam still remains horizontal the balance is just. The sensibility of a balance is tested by making a small addition to one of the pans.

Questions.

1. A beam, 40 feet long, is supported horizontally at its two extremities. Determine the additional pressure on the supports when a weight of 2,400 lbs. is hung at a point 10 feet distant from one end. *Ans.*: 600 lbs. and 1,800 lbs.

2. At two points, A and B, in the same horizontal line, the extremities of a cord are fastened, and a weight of 10 lbs. is suspended from the middle point C of the cord. Find the tensions in the portions of the cord CA and CB, it being observed that the angle ACB is a right angle. *Ans.*: $10 \div \sqrt{2}$.

3. In fig. 55 the weights P and Q are 2 lbs. and 3 lbs. respectively; if the length CX be 6 inches, determine the perpendicular CY when the disc is at rest. *Ans.*: 4 inches.

4. The arms of a false balance are 1 foot and 1·03 foot in length; what will be the apparent weight of 10 lbs. of goods, weighed with this balance, when suspended from the longer arm. *Ans.*: 10·3 lbs.

5. If the goods be suspended from the shorter arm, find their apparent weight. *Ans.*: $10 \div 1{\cdot}03$.

6. If the beam of a balance be not horizontal, show that if want of horizontality arise from an inequality in the weights of the scale pans, the balance may be corrected by putting a weight into the lighter of the two pans, but if it arise from a difference of length of the arms, the balance cannot be so corrected.

7. Find the true weight of a body which is found to weigh 8 ounces and 9 ounces according as it is placed in one or the other scale pan of a false balance. *Ans.*: $6\sqrt{2}$.

CHAPTER VIII.

THE INCLINED PLANE.

§ 54. **The Inclined Plane.**—A plane which is inclined to the plane of the horizon is called an inclined plane; and the angle which it makes with the plane of the horizon is called the inclination of the plane.

When a weight is placed upon a *horizontal* plane it will remain at rest. In this case the effect of the weight of the body is only to press against the plane; and this tendency is counteracted by the reaction of the plane. But when a body is placed upon a smooth *inclined* plane, the circumstances are different. The reaction of the plane cannot then entirely support the body, and consequently the body will slide down. As a matter of fact, it is impossible to make a perfectly *smooth* plane, and the roughness of the plane will be sufficient to prevent a body from sliding down it unless the inclination be considerable. For the present we shall, however, consider that the plane is perfectly smooth; and in the next chapter it will be shown how the results are modified by what is called friction.

In order to support a heavy body which rests on an inclined plane, H F (fig. 58), a certain force may be applied to the body in a direction parallel to the plane. It is required to determine the magnitude of that force.

FIG. 58.

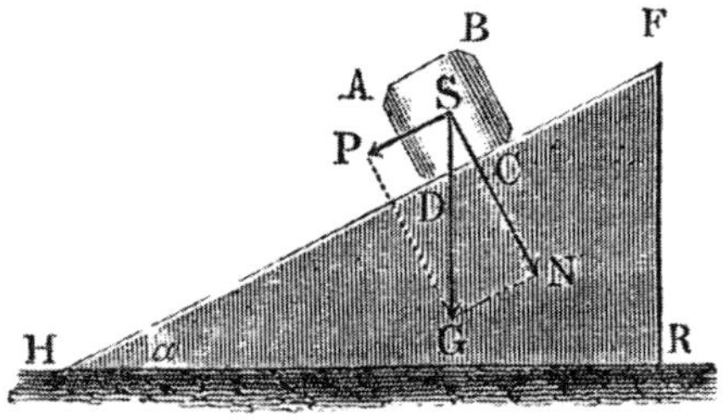

Let S be the centre of gravity of the body, and let the vertical line S G denote the magnitude of the force of gravity upon the body. Through S draw S P parallel to H F, and S N perpendicular thereto. Draw through G lines G N and G P parallel to H F and S N. Then, by the principle of the parallelogram of force, we may replace the force S G by the two forces S P and S N. Now, S N is perpendicular to the plane; and its effect is merely to press the body towards the plane, which tendency is counteracted by the reaction of the plane. The component S P tends, however, to drag the body down the plane, and unless it be opposed by an equal and opposite force the body will descend. The required force is therefore equal and opposite to S P.

It is easy to show that the triangle P S G is similar to H F R, and therefore the sides about the equal angles are proportional.

We thus have

S P : S G :: F R : F H,

whence the following general statement :

If a body resting on an inclined plane is in equilibrium under the influence of a force applied parallel to the plane, then this force must bear to the weight of the body the same ratio which the height of the plane bears to its length.

FIG. 59.

Example.—The length of an inclined plane is 8 feet and its height is 2 feet. Find the power applied parallel to the plane which will be sufficient to sustain a load of 100 lbs.

Since the height is in this case one-fourth of the length, it is obvious that the power will be one-fourth of the load, that is to say, 25 lbs.

The principle of the inclined plane may be illustrated by the well-known contrivance used in lowering a hogshead down a slope or a flight of steps. (See fig. 59.) Two ropes are attached by one of their extremities to a piece of wood, which is placed at the top of the inclined plane. These ropes go down the plane, are passed around the hogshead in the way shown in the figure, and the other extremities are held by the two men. By slowly letting out the rope the men are able to lower down the weighty hogshead as gently as can be desired.

If the two ropes are passed round the hogshead at equal distances from its extremities, the two men will have to exert the same force. It is also to be observed that the two parts of the cord which are fastened to the piece of wood are acted upon by forces equal to those acting in the two parts held by the men. Thus the hogshead is really acted upon by four parallel forces. The force with which the hogshead is urged down the plane is equally divided among the four parts of the rope which sustain it.

If the height of the plane be one-half its length, then the total force necessary to sustain the hogshead will be half the weight of the hogshead. It therefore follows that the tension in each part of each of the ropes will be one-eighth the weight of the hogshead. Each of the men has therefore only to exert a force which is equal to one-eighth of the weight of the hogshead in order to have perfect control over its movements.

§ 55. **Two Inclined Planes.**—Two weights, A and

B (fig. 60), are attached to the extremities of a cord, which passes over a pulley C, at the vertex of two inclined planes. It is required to find the relation between the weights, in order that they shall balance each other.

FIG. 60.

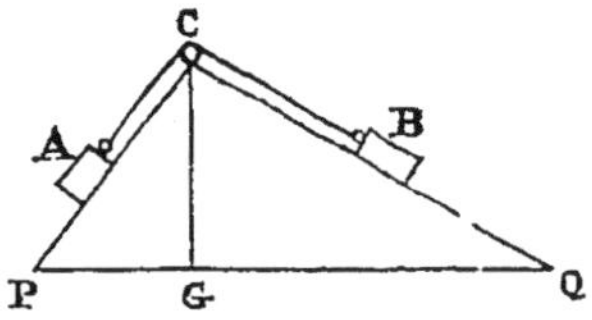

In this case the immediate support of each weight is the tension of the cord which passes over the pulley. We therefore have the following proportion:

Tension of cord : weight of A :: C G : C P,

or

Tension of cord . C G :: weight of A : C P.

But from the other weight we have also

Tension of cord : C G :: weight of B : C Q.

Whence obviously

Weight of A : C P :: weight of B : C Q.

It is therefore obvious that each of the weights is proportional to the length of the plane on which it rests.

FIG. 61.

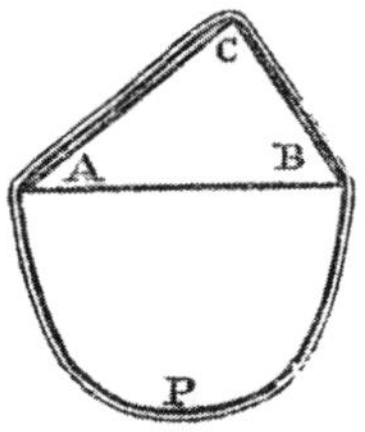

This can be shown in a different way by a somewhat curious method of reasoning. In fig. 61 let A B C be a triangle, of which the side A B is horizontal. Suppose that a chain is hung round the triangle, so that while it is in contact with the sides at A C and B C it hangs down loosely at P. It is evident that since A B is horizontal, the part A P B of the chain is symmetrical, and therefore the tension of the chain at A is equal to that at B. What, then, is the tension at C? It must consist of the tension at A augmented by whatever tension is produced by the weight of the portion A C, or it must consist of the tension at B augmented by whatever tension is produced by the portion B C. Since these must be equal to each other, it follows that the tension due to the part of the chain A C must be equal to that due to the part of the chain B C. But the weights of the chain are respectively proportional to the lengths of the sides on which they rest; and thus it appears that any two weights on the sides connected by a string over a pulley at the top will remain in equilibrium, provided that the magnitude of each weight is proportional to the length of the plane upon which it rests.

FIG. 62.

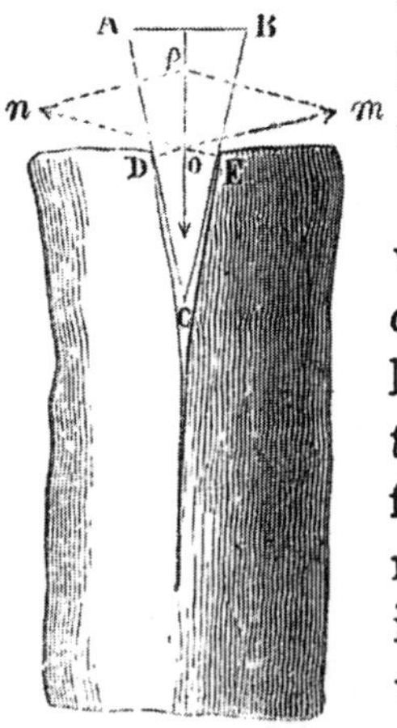

§ 56. **The Wedge.**—There is a very useful application of the inclined plane which enables a very large resistance to be overcome by the action of a comparatively small force. The appliance to which we refer is known as the wedge, and in one form or another is of very wide utility. The particular form of the instrument which we shall describe is the wedge used in splitting timber. This consists of a piece of iron of a form which will be at once understood from fig. 62. The wedge is urged into the timber by the blows of a hammer applied on the face, A B, and the edges of the timber at D and E are forcibly driven asunder. We shall not at present enter into the question of the peculiar efficiency which the blow of a hammer possesses for such a purpose as that now under consideration. It will be sufficient here to observe that the effect of the blow is to produce a pressure upon the face, A B; and it will be for us to consider how that pressure is transformed into the forces at D and E.

For this purpose we observe that there are three forces acting upon the wedge, which make equilibrium. These are, firstly, the pressure on the face, A B; and secondly, the pressures exerted on A C and B C respectively by the sides of the cleft at D and E. These forces are perpendicular to the faces of the wedge, and act along the directions D *m* and E *n*. If we take *o p* to represent the pressure on the face, A B, and if through *p* we draw parallels *p m* and *p n* to the lines *n* E and *m* D, then the pressures at D and E will be represented by *m o* and *n o* respectively.

We thus see that the ratio between the pressure exerted on A B and the forces applied at D and E is the same as the ratio of the lines *o p* and *o n*. The more acute the angle at C the greater is the ratio in which the pressures at D and E exceed the pressure on A B.

QUESTIONS.

1. What force is necessary to sustain a weight of 10 lbs. on a plane inclined to the horizon at an angle of 30°, the force acting horizontally?

FIG. 62 (*a*).

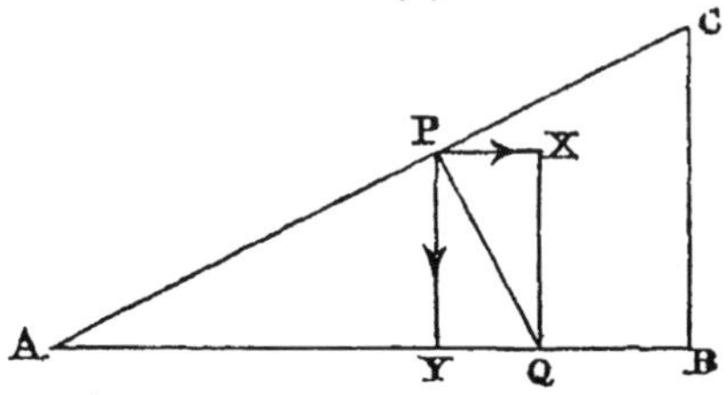

Ans.: Draw through P, fig. 62 (*a*), a line, PQ, perpendicular to AC; draw QX, parallel to BC, and also PX and PY, parallel to BA and BC respectively. Then, if the weight of the body be represented by PY, the horizontal force required is PX. It is obvious that the triangle PQY is similar to ACB, and therefore YQ or PX : PY :: CB : AB. It can be easily shown from geometry that, when the angle at A is 30°, BC : AB :: 1 : $\sqrt{3}$, and, therefore, the required force, PX, is $10 \div \sqrt{3}$.

2. What force is necessary in the last question if the force acts vertically? *Ans.:* 10 lbs.

3. What weight could be supported, on a plane inclined at an angle of 30° to the horizon, by a man who was just strong enough to lift vertically a weight of 300 lbs.? *Ans.:* 600 lbs.

4. Two equal weights of 10 lbs. are supported on two opposite inclined planes by a cord which is parallel to the base of the planes. If the planes are inclined to the horizon at an angle of 45°, determine the tension of the cord. *Ans.:* 10 lbs.

5. If a wedge be 12 inches long and one inch wide at the base, find the force exerted on the timber by a pressure of 250 lbs. on the end of the wedge. *Ans.*: 3000 lbs.

CHAPTER IX.

FRICTION.

§ 57. **Friction.**—In what has been stated in the last chapter it has frequently been necessary to make assumptions which cannot be realised in practice. It has been supposed that the surfaces on which the experiments have been made were *perfectly* smooth. Now, in reality, this state of things can only be approximated to. All real surfaces are in some degree *rough*, the effect of the most careful smoothing or polishing being only to lessen the roughness. The consequence is that, when bodies are in contact, they do not move over each other with perfect freedom; their relative motion is always more or less resisted, and the resistance met with under these circumstances is called *friction*.

This may be illustrated by the following considerations. A body resting on a horizontal plane is acted upon by gravity, which produces a vertical force, acting downwards at the point G, fig. 63. Now, if the plane were perfectly smooth the minutest force acting at F would be sufficient to move the body along the plane. If, however, the weight were an irregular mass of stone, and the horizontal plane was the rough floor of a quarry, it is evident that a very large force would be required to be applied at F in order to move the block of stone along. The labour will, however, be greatly facilitated by placing rollers beneath the block, for

FIG. 63.

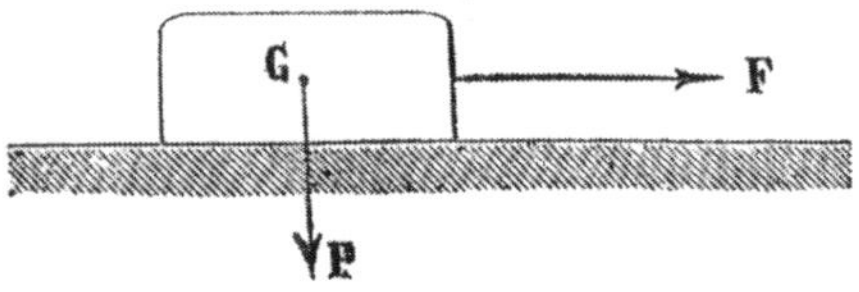

then it is found that the force F which is necessary to move the block is considerably lessened. Still more will the transit of the block be facilitated when it has been raised upon a truck of which the wheels run along a tramway. One horse will then be able to draw a block along with ease, which he could not stir as long as it lay on the quarry floor. If, finally, the stones are placed in an ordinary railway waggon, they are moved with even greater facility. It is then found that a force of about 10 lbs. is sufficient to draw a weight of one ton along the level. Thus the rude contrivance of the roller and the more finished contrivance of the wheel are merely devices for diminishing the force of friction; and the wheels themselves are oiled for the purpose of diminishing the friction at their axles.

It therefore appears that the weight G, fig. 63, will require a certain force F to be applied to it in order to draw it along the plane.

The magnitude of this force will depend upon two circumstances, viz., the magnitude of the weight G, and the character of the two surfaces which are in contact. It will be obvious, from the illustration already given, that the rougher the surfaces in contact the greater must be the force F. It is equally plain that, the surfaces in contact remaining the same, a greater weight at G will require a greater force at F.

§ 58. **Experiments on Friction.**—The investigation of the relation between the power F and the weight G, when the surfaces in contact remain the same, has been the subject of numerous experiments. The apparatus employed for this purpose is represented in fig. 64. On a firm horizontal table, B, a slab of one of the two surfaces to be experimented upon is secured; on this is laid a slab of the other substance; to this slab a cord is attached, which passes horizontally to a pulley, C (firmly fixed to the table at B), and then descends vertically to support a tray at D for the purpose of carrying weights. Another tray, A, is placed upon the slab to which the cord is fastened, and by putting suitable weights into this tray any required pressure can be produced between the two surfaces in contact.

The experiments are then made in the following

FIG. 64.

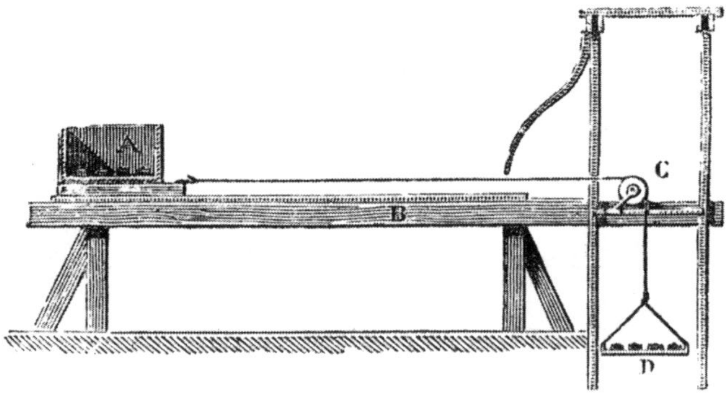

manner. A certain load is placed upon the tray at A, and then weights are placed upon D, until the tension of the cord is sufficient to draw the slab uniformly along, bearing, of course, the weighted tray. In a second experiment the load at A is altered, and the corresponding weight required at D is determined. When a number of experiments have been made, it is found that there is an exceedingly simple relation between the force with which the two slabs are pressed together, and the force which will just draw the upper slab along. This relation may be expressed as follows, and is often called the first law of friction :—

The force of friction is proportional to the mutual pressure between the two bodies.

For example, let us suppose that it has been observed that, when the pressure between the two surfaces is 10 lbs., the friction is 4 lbs.; then, when the pressure is 20 lbs., the friction will be 8 lbs., and so on.

It must, however, be clearly understood that this law is merely an approximation. A considerable degree of uncertainty always attends experiments of this nature. All that we can assert positively is that the observed values of the friction are fairly represented by the supposition that they are proportional to the load.

The following are a few values of the constant ratio between the friction and the load, deduced from experiment:

Wood upon wood .	. 0·50
Metal upon wood .	. 0·60
Metal upon metal .	. 0·18.

For example, a smooth block of iron weighing 100 lbs. will be drawn by a force of 60 lbs. along a smooth surface of wood, or by a force of 18 lbs. along a smooth surface of iron.

The ratios here given are much reduced by the presence of a lubricating material between the surfaces in contact. For example, if olive oil were placed between the two surfaces of iron in the case just mentioned, a force of 12 lbs. would be sufficient.

We have hitherto spoken only of the effect of the total pressure between the surfaces; it now remains to consider the effect of the area of the surfaces which are in contact. For this, again, we must resort to experiment, and may use for the purpose the apparatus employed before, only being provided with moveable slabs of different sizes. It is found by experiment that, so long as the total pressure between the slabs is constant, the area of the surface of contact of the slabs is without effect. To make this clear, let us take a special case. Suppose that the two slabs are each composed of wood, and that they are pressed together by a force of 10 lbs. Then the friction will be found to be 5 lbs., and this will be the same whether the surface of contact has an area of 10 square inches, or of 20 or 30.

A little reflection will show that this result might have been anticipated from the first law of friction already explained. When the surface of contact has an area of 10 square inches, and the pressure between the two slabs is 10 lbs., then the pressure between the two surfaces is 1 lb. per square inch; the amount of friction is $\frac{1}{2}$ lb. per square inch, and the total friction is 5 lbs.

But when the area of contact is 20 square inches, while the load is still 10 lbs., the pressure per square inch is $\frac{1}{2}$ lb., and by the first law the friction per square inch will be $\frac{1}{4}$ lb.; but, as the total number of square inches is 20, the total friction will still be 5 lbs. as before.

The second law of friction is expressed in the following terms :—

The force of friction between two bodies is independent of the size of the surface in contact.

§ 59. **Friction on the Inclined Plane.**—The effect

of friction on the equilibrium of a body resting upon an inclined plane may be studied experimentally by the apparatus represented in fig. 65. A B is an horizontal plane to which another plane A C is attached

FIG. 65.

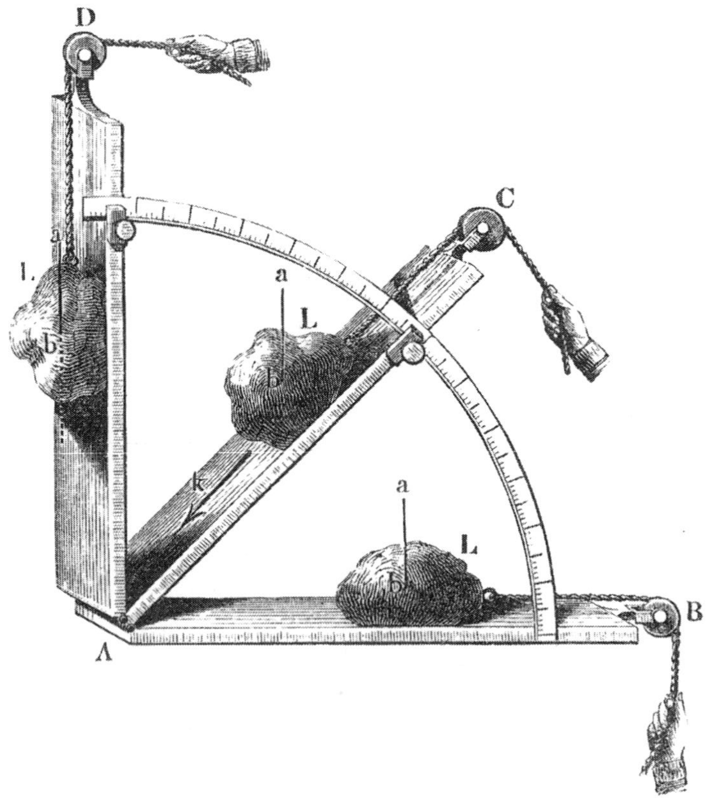

by hinges. This moveable plane can be clamped at any required inclination to the horizon by means of the divided arc. At B and C pulleys are attached, over which a cord passes which can be attached to the body L.

In the case of the body L in contact with the moveable plane in the vertical position, it is clear that there is no pressure of the body against the plane, and that, consequently, there is no friction. In order to maintain the body in this position a force must be exerted which is simply equal to the weight of the body. In the case of the body L resting upon the horizontal plane A B, a force exactly equal to the friction between the body and the plane must be exerted in order to move the body.

Let us now suppose that the inclination of the plane is gradually increased. While the inclination is still small, the body remains as steady as when the plane was horizontal, but at length a certain inclination is reached, which is represented by the position A C in the figure. At this inclination, which is called the *angle of friction*, the body is just on the point of descending the plane under the action of gravity. If the inclination of the plane be at all increased, then the body will descend unless held back by the cord passing over the pulley at C. As the inclination of the plane is increased, so the tension of the cord necessary to hold the body back will increase, until it amounts to the full weight of the body, when the moveable plane has attained the vertical position.

Both the laws of friction can be illustrated and confirmed by the use of this apparatus. It is observed that so long as the character of the two surfaces which are in contact remains the same, the angle of friction remains the same whatever be the weight of the body pressing upon it. Now the friction is equivalent to a force acting along the inclined plane, and we have seen that, when a body is maintained in equilibrium by the action of such a force, the intensity of the force must bear to the weight of the body the same ratio which the height of the plane bears to its length. The length of this inclined plane is always constant, and its height is constant when it is placed at the same inclination. It therefore follows that the ratio of the friction to the weight of the body is constant, which is simply the first law of friction.

It is also observed that so long as the *character* of the surfaces in contact remains unaltered, the angle of inclination is not dependent upon the area of the surfaces, and therefore the second law of friction is demonstrated.

It is hardly possible to over-estimate the importance of the force of friction. We often recognise it as a destroyer or impeder of motion, and as a source of loss and inconvenience. But, on the other hand, friction is often indirectly the means of producing motion, and of this we have a splendid example in the locomotive engine. The engine being very heavy the wheels are pressed closely to the rails; there is friction enough to prevent the wheels from slipping; consequently, when the engines force the wheels to turn round they must roll onwards. We have already seen that the friction of iron upon iron is about 0·18 of the pressure. Suppose a locomotive engine to weigh 30 tons, and the share of the weight borne by the driving wheels be 10 tons, the friction between the driving wheels and the rails will then be 1·8 tons. This is the greatest force the engine can exert upon a level line. A force of 10 lbs. for every ton weight of the train is known to be sufficient to sustain the motion, consequently the engine we have been considering should be capable of drawing along the level a load of

$$1·8 \times 2240 \div 10 = 403 \text{ tons.}$$

But we need not go to the steam engine to learn the use of friction. We could not exist without it. In the first place we could not move about, for walking is only possible on account of the friction between the soles of our shoes and the ground. Nor if we were once in motion could we stop without coming into collision with some other object, or grasping something to hold on by. Objects could only be handled with difficulty, nails would not remain in wood, and screws would be equally useless. Friction is, in fact, as far we are concerned, quite as essential to the economy of nature as the law of gravitation.

§ 60. **Applications of Friction.**—Among the numerous instances in which friction is made use of in mechanical appliances we may mention the brake used in stopping a railway train. The essential parts of this contrivance are shown in fig. 66. M is a handle turning a screw in a fixed nut V; this screw is attached by a rod to the bell crank lever *a* O *b*, which turns round the point O; the

FIG. 66.

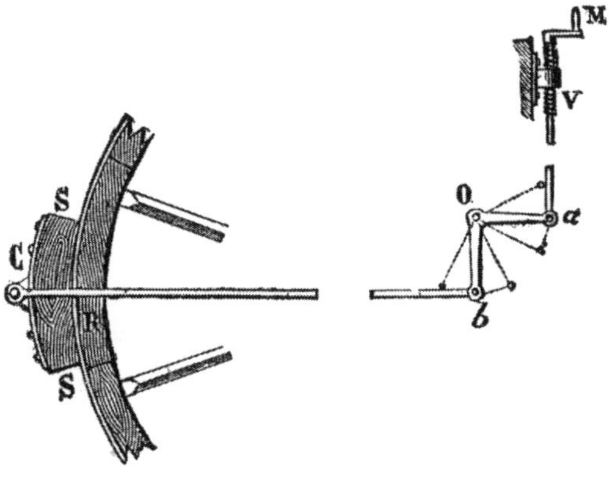

other extremity of the bell-crank lever is attached to the rod *b* c, which acts upon the brake s s. When the train is in motion the wheel R is revolving, and the brake s s is quite close to the wheel, though not pressed against it. When it is desired to stop the train, the handle M is turned, the point *a* is raised, the point *b* is drawn to the right, and the brake s s is forced against the edge of the wheel.

Fig. 67.

§ 61. **Use of Belts in Machinery.**—A very extensive application of friction is met with in the belts by which it is customary to distribute the power of a steam engine from the shaft which traverses the mill to the different machines which are worked thereby.

A simple case of this is shown in fig. 67, in which a grindstone is receiving its motion from a shaft A B. On the shaft is fixed a drum, which revolves with it, and over this passes a leather belt which embraces a similar drum on the axis of the grindstone. When the first drum revolves the friction between it and the belt is sufficiently great to force the belt to move too, this forces round the drum attached to the grindstone and thus the grindstone revolves.

§ 62. **Coil Friction.**—A very convenient means of producing a very great resisting force is afforded by the friction between a *fixed* cylinder and a rope which is coiled round it. In fig. 68 is shown a cylinder A B, which is fixed securely on a beam D E. A weight P is attached to a rope which after being once passed round the cylinder is held in the hand at Q. It is found that so far as the *descent* is concerned, the weight P can be held under complete control by the exertion of a comparatively small force at Q. If the cylinder be ordinarily rough the force which must be exerted at Q, in order that P may be sustained, is only about one-twentieth

FIG. 68.

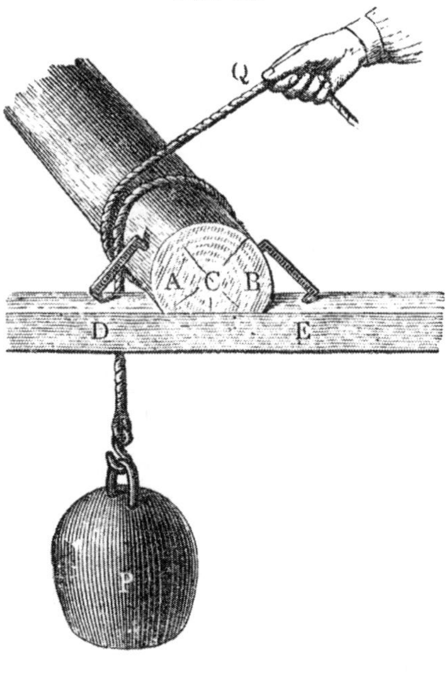

part of the load P. Thus, if P weigh 200 lbs., it can be securely held up by a force of 10 lbs. at Q.

The efficiency of this contrivance can be enormously increased by increasing the number of coils made by the rope around the cylinder. Assuming as before that for one coil the load is twenty times the power; then for two coils the load is $20 \times 20 = 400$ times the power; for three coils, the load is $20 \times 20 \times 20 = 8000$ times the power; and so on.

This principle has a very useful mechanical application in the operation of bringing a steamer to rest when she comes alongside the quay. One end of a strong rope is made fast to the vessel, while the other end is coiled two or three times round a firmly fixed post on shore. A man holding the free end of the rope can thus exert very great power over the vessel, and can bring her to rest at the desired moment.

§ 63. **Wheels.**—Having mentioned a few contrivances in which the force of friction is utilised, we have now to point out how it may be partially evaded when its influence is deleterious. The wheel is one of the most simple and effective contrivances for overcoming friction. A sleigh is a very admirable vehicle on a smooth surface, such as ice, but it is totally unadapted for use on common roads, the reason being that the amount of friction between the sleigh and the road is so great that to move the sleigh the horse would have to exert a force which would be very great compared with the load he was drawing. But a vehicle properly mounted on wheels moves with the greatest ease along the road, for the circumference of the wheel does not slide, and consequently there is no friction between the wheel and the road. The wheel, however, turns on its axle, therefore there is sliding, and consequently friction, at the axle, but the axle and the wheel are very perfectly fitted to each other, and the surfaces are lubricated with oil, so that the friction is comparatively small.

With large wheels the amount of friction on the axle is less than with small wheels. Other advantages of large wheels are that they do not sink much into depressions in the roads, and that they have also an increased facility in surmounting the innumerable small obstacles from which even the best road is not free.

QUESTIONS.

1. State the laws of friction.

2. A body, weighing 12 lbs., is just on the point of slipping on a rough inclined plane which rises 2 feet in a length of 8 feet. What force would draw the body along the same plane if horizontal? *Ans.:* 3 lbs.

3. Show that when two smooth surfaces are at rest in contact with each other, the pressure between them must be perpendicular to the two surfaces. Is this true if the surfaces be rough?

4. A weight of 20 lbs. at P is *supported* at Q (fig. 68) by a force of 1 lb. What force would have to be exerted at Q in order to *raise* 20 lbs. suspended at P? *Ans.:* 400 lbs.

5. Describe the application of friction in the case of belts used for transmitting motion to machinery.

CHAPTER X.

ENERGY.

§ 64. **Work.**—If a man lifts a weight which was previously lying on the floor, and places it upon the table, he is said to do *work* in the sense in which the word is used in mechanics. The amount of *work done* depends both upon the magnitude of the weight and also upon the height of the table. It is clear that we cannot tell the amount of exertion that will be required to raise a given weight until we know the height through which it is to be lifted, nor will it be sufficient to know the height unless we also know the weight.

We may adopt as the *unit of work* the small quantity of exertion which is necessary to raise a weight of one pound through a height of one foot: this is often called the *foot-pound.*

If a weight of 1 lb. has to be raised through a height of 2 feet, or a weight of 2 lbs. through a height of 1 foot, it will be necessary to expend twice as much energy as would have raised a weight of 1 lb. through 1 foot, that is 2 foot-pounds.

If a weight of 5 lbs. had to be raised from the floor up to a stool 3 feet high, how many units of work would be required? To raise 5 lbs. through 1 foot requires 5 foot-pounds, and the process must be again repeated twice before the weight arrives at the top of the stool. For the whole operation 15 foot-pounds will therefore be necessary.

If 100 lbs. be raised through 20 feet, 100 foot-pounds of work is required for the first foot, the same for the second, third, &c., up to the twentieth, making a total of 2000 foot-pounds.

These examples will be sufficient to illustrate the following general statement:

The amount of work done in raising a given weight through a given height is expressed in foot-pounds by multiplying the number of feet in the given height by the number of pounds in the given weight.

We may practically illustrate this principle by solving the following question: Would you have to expend as much work in lifting a weight of 50 lbs. to a height of 15 feet as would be required to raise a weight of 40 lbs. to a height of 20 feet? In the first case the foot-pounds of work done are $50 \times 15 = 750$, in the second case they are $40 \times 20 = 800$; the answer is therefore negative.

§ 65. **Energy.**—The power of doing work is called *energy*, and whatever is capable of doing work is said to possess energy. We shall illustrate this conception by a few simple examples.

In winding a clock I am conscious of exerting pressure with my hand in order to turn the key. I am therefore doing work, which in this case is expended in raising the weight by which the movement of the works in the clock is sustained. Now when the weight has been raised it is capable of doing work, and the particular work which it is called upon to do is that of keeping the works of the clock in motion against the resistances to which they are exposed. The raised weight of the clock is therefore capable of doing work; it may therefore be said to possess energy. Observe, however, that as the weight only possesses energy when it is raised, the energy is simply due to the *position* of the weight. In such a case we say that the energy is *statical.* Had the clock been driven by a spring instead of a weight, then in winding the clock I do work as in the former case, only this work is expended upon overcoming the resistance of the spring to being wound. When the winding is completed the spring may be said to possess statical energy, for it possesses something which gives it the power of doing work until the time comes for winding the clock again.

Gunpowder may be likened in some degree to a highly compressed spring. It is in fact largely charged with statical energy which can be suddenly transformed into work. When a rifle is discharged, the statical energy of the gunpowder is expended in giving velocity to the bullet, but the energy originally in the powder is only transformed, it is not lost. A little consideration will show that in this case the bullet actually carries away with it some of the energy originally possessed by the gunpowder. Had the rifle been pointed vertically upwards, it is plain that the bullet would have gone up vertically also. Therefore the bullet must have possessed energy when it left the gun, because it was able to raise a weight, *i.e.* its own weight.

Without supposing the bullet to be shot vertically upwards, we can still show that it possesses energy when it leaves the gun. Imagine a deep hole in the ground, at the bottom of which is a small weight to which a cord is attached, which, passing over a pulley at the top of the hole, is attached to the bullet in the rifle. Then, even if the rifle were fired horizontally, it is obvious that (if the cord did not break) the weight in the hole would be lifted; therefore the moving bullet has done work, and therefore it must possess energy.

We are thus led to the conception that a body in motion possesses energy, and this is often called *kinetic* energy, to distinguish it from *statical* energy already considered.

In the case of the rifle bullet fired vertically upwards, we have a good illustration of the change of kinetic energy into statical energy, as well as the converse operation. When the bullet leaves the mouth of the rifle (which for simplicity we shall consider to be at the surface of the earth) all its energy is kinetic; as the bullet ascends its velocity diminishes, and there is a corresponding decrease in the kinetic energy. On the other hand, it is gradually acquiring statical energy, for, as its distance from the earth increases it would be capable of doing more and more work by merely being allowed to descend. At length a moment arrives when the upward velocity ceases, the bullet is then for an instant at rest, and its kinetic energy is then zero. But now the statical energy is at its maximum, for the bullet would be able to do work while descending through the entire distance which it has attained. If, however, the bullet be allowed to descend freely; its statical energy will gradually become transformed into kinetic energy, and when the bullet finally reaches the ground it will have regained all its original kinetic energy (saving what has been lost by the resistance of the air), while it will have ceased to possess any statical energy.

In the range of our experience we do not find any case in which it can be said that energy is created. It may be thought at the first glance that by the aid of certain contrivances (for example, steam-engines) we are enabled to create energy, but this is not really so. Energy subsists in a vast number of forms, and by our contrivances we are able to turn one form of energy into another form, but this is a very different matter from creating energy. It is of the utmost importance that this be clearly understood, and therefore we shall consider several cases, in which, by the aid of mechanical contrivances, we are enabled to transform pre-existing energy into a useful form.

Take first, as the most simple instance, the case of a flour-mill driven by an ordinary overshot water-wheel. The wheat is crushed between heavy stones, and to force these stones to do their work a quantity of energy is required. This energy is communicated by the shafting and wheel-work from the water-wheel to the mill-stones. But the water-wheel does not create the energy; it merely seizes on a portion of the statical energy of the water, and transforms it into a convenient shape. The conditions under which a water-wheel will work are found in the presence of a mill-pond, and a channel at a lower level than the pond, by which the water can be carried away after it is used. Thus the water in the pond has the statical energy, which is due jointly to its weight and to its height above the channel. The quantity of work which can be accomplished inside the mill can never exceed the number of foot-pounds, which is found by multiplying the number of pounds of water used into the number of feet through which the water has descended. As a matter of fact the quantity of useful work which can actually be done falls considerably short of the theoretical efficiency of the wheel. This is due partly to the friction in the machinery, and partly to the imperfect manner in which the water-wheel itself acts. It is, however, believed that if all the statical energy of the water could be utilised, then it would be precisely equal to the quantity of work done. It will thus be seen that the water-wheel can in no sense be said to be a creator of energy.

Let us push the enquiry back a step further, and enquire how it is that the water has acquired the statical energy by which it was enabled to drive the mill. The water originally must have fallen in the form of rain, and is descending by a river on its way to the sea, when it is intercepted by the dam which conveys it to the mill-pond. But how came the water to be raised into the air, so as to form rain? There can be no doubt that the water was first of all drawn from the sea, whence it ascended in vapour into the atmosphere. This process was induced by *the heat of the sun.* We are thus led to the conception that the rays of the sun convey energy to the earth, and that consequently the work done in grinding corn in the mill is due to the energy which the sun possesses in virtue of its enormous temperature.

The horse which conveys the corn to the mill, as well as the labourer who carries it in, are each performing work, but neither the horse nor the man can be said to create the energy which enables them to do work. The energy of the horse is derived from the hay and oats on which he is fed; but whence do the meadow and the oat-field derive the energy which is certainly found in their crops. Here again we trace the source of the energy to the sun, under whose influence the crops are grown; but it would lead us too far to endeavour to explain the process by which the growing plant seizes on the energy which it finds in the sun's rays, and consolidates that energy within its foliage and fruit.

Finally, let us suppose that the miller calls in the aid of a steam-engine to turn his mill-stones. Can he by this means create energy? We must in this case look to the coal with which his boilers are supplied as the real source of the energy. The steam-engine is in fact merely a convenient method of transforming the energy latent in coal into a useable shape.

But where did the coal derive its energy? It seems certain that in ancient days the earth was furnished with a luxuriant vegetation, and that portions of this vegetation have been preserved by peculiar circumstances in the form of coal. Here, again, we see that the rays of the sun are the source of energy by which our steam-engines are moved.

§ 66. **The Pulley-block.**—It may, however, be thought that by the use of a pair of pulley-blocks, or other means whereby a small force can be made to overcome a large one, it would be possible to transform a given quantity of energy into a larger quantity. This is not the case.

In fig. 69 is shown an arrangement whereby a

weight P is able to balance a larger weight W; the relation between the weights P and W is to be deduced from the following considerations: The cord attached to P passes over a pulley in the upper block, and then descends to a pulley at the lower block, thence up again to the upper block, and so on, until three pulleys in the upper block and three in the lower block are embraced by the cord, the end of which is finally secured to the bottom of the upper block.

FIG. 69.

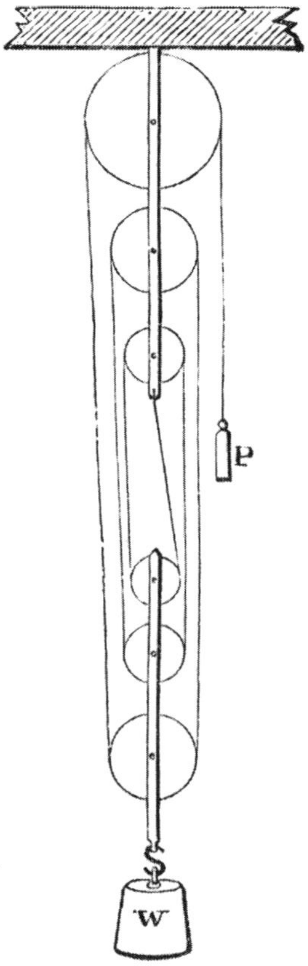

The tension of the cord will be unaltered by passing over the pulleys *if there is no friction*, and therefore the tension of the cord throughout its entire length will be simply equal to the weight P.

It will be observed that the weight W (and that of the lower block) are supported by six different parts of the cord, and since each of these is strained by the weight P, and since they are all practically parallel to each other, their resultant will be 6 P. Hence the weight W and that of the lower block are together six times the weight of P.

It thus appears that by this arrangement a weight P is able to balance a weight which is six times as great as itself, and that therefore the slightest increase to

P would actually cause it to descend and raise W. But can energy be created by such an operation? Let us suppose that P is 1 lb., and that therefore the load it supports by the aid of the pulley is 6 lbs. Now if P descend a distance of 6 feet it will do six units of work. But as there are six portions of the cord between the two pulleys, then when P has descended six feet and pulled out six feet of the cord, each of the portions of cord between the two pulley-blocks will be shortened one foot, and therefore the weight W will be raised 1 foot. It therefore appears that when six units of work are done by P, a weight of 6 lbs. will have been raised 1 foot; *i.e.* six units of work will have been done against W. It follows that the six units of work applied directly to W would have accomplished just as much work upon it as when they were applied by the intervention of the pulley-block.

FIG. 70.

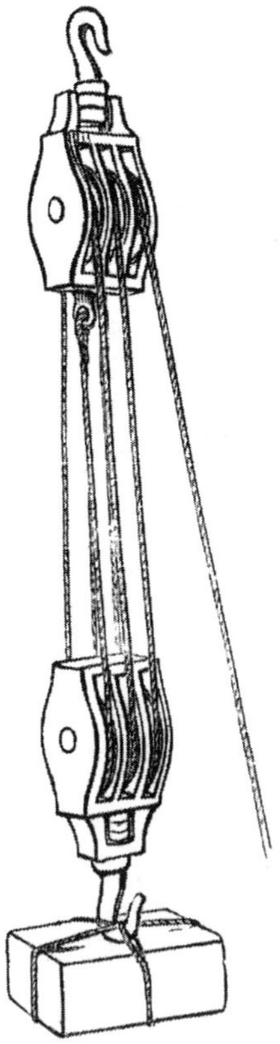

The convenience of the pulley-block arises from the circumstance that though it cannot increase energy, yet it enables a large resistance to be overcome. Suppose, for example, that a weight of 500 lbs. has to be raised a certain height; a man could not lift such a weight without some such aid as a pulley-block could afford him; but the total quantity of energy which he must expend to accomplish the work with the pulley-block will be quite as much as would be sufficient to raise the weight if he could divide it into the portions of 50 lbs. and then lift each of them separately by his hands alone.

The efficiency of the pulley-block is very considerably impaired by the action of friction. Instead of the power being only one-sixth of the load, it is found that in a pair of ordinary three-sheave pulley-blocks a power of one fourth of the load is required. Under these circumstances there is positively a loss of energy by the use of the pulley-block. The most convenient practical form of the instrument is shown in fig. 70.

QUESTIONS.

1. Define a foot-pound of work.

2. How many foot-pounds will be expended in raising a weight of 90 lbs. through a height of 10 feet? *Ans.:* 900.

3. How is it shown that a body, when in motion, possesses energy?

4. Distinguish between statical energy and kinetic energy. How is it shown that energy is not created by a water-wheel?

5. What is the origin of the energy by which a water-wheel is set in motion?

6. Is energy *created* by a steam-engine? If not, from what source does it come?

7. A steam-engine of one horse power is capable of performing 33,000 foot-pounds of work per minute. How many sacks, each weighing 14 stone, will such an engine be able to lift 50 feet in eight hours? *Ans.:* 1616.

8. Describe the three-sheave pulley-block, and explain its use.

CHAPTER XI.

PRACTICAL ILLUSTRATIONS OF MECHANICAL PRINCIPLES.

§ 67. **The Windlass.**—A cylinder A (fig. 71) has a handle attached to it at M; and round the cylinder a rope is wound from which hangs the bucket P. The cylinder is free to revolve in its bearings, and when the handle is turned the cylinder revolves and winds up the rope. This well-known contrivance is called a Windlass, and is often used for raising water from a well. We have to ascertain the relation between the power F, which is applied at the handle M, and the load P which is raised in the bucket.

FIG. 71.

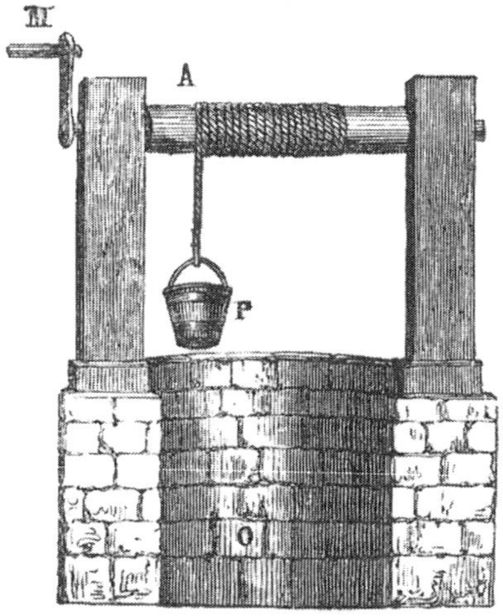

We shall suppose that the power is applied in a direction which is perpendicular to the arm of the handle, the relation between the power and the load can then be determined by the principle of the bent lever. Fig. 72 represents a section of the cylinder; the bucket, of which the weight is P, hangs vertically from A, while the power F is applied perpendicularly to O B.

FIG. 72.

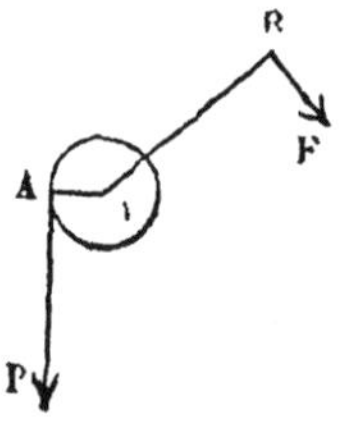

For equilibrium we must have the following condition:

The power must be to the load in the same ratio as the radius of the cylinder to the radius of the circle described by the handle.

§ 68. **The Capstan.**—This very useful appliance is represented in fig. 73. A vertical cylinder is mounted so that it can revolve in fixed bearings. This cylinder is made to revolve by the pressure of one or more men against the four arms which stand out from the

FIG. 73.

top of the cylinder in a direction perpendicular to its length. The power produced is transmitted by means of a rope which receives a few turns round the cylinder in the way shown in the figure. The man who is sitting down holds the free end of the rope in his hands and coils it up as it is gradually wound in. In order to understand why the rope does not slip, reference must be made to § 62. In this case, since the rope makes two complete coils, the man who is holding the rope need only exert upon it a force which is the 1 ÷ 400th part of that exerted on the load to which the other end of the rope is attached. The relation between the power applied at the handle and the load overcome by the rope is easily found from the principle of the bent lever, and we have the following result:

The power applied to the arm of a capstan bears to the load the same ratio which the radius of the cylinder bears to the length of the arm.

Suppose, for example, that the radius of the cylinder of the capstan was 7 inches, and that each of the arms was 7 feet, then when one man is working as represented in the figure, twelve times the force which he exerts will be transmitted along the rope. If four men are working the capstan, then the force transmitted along the rope will be 48 times the force exerted by each of the men.

The capstan is used when a great force is required to be transmitted horizontally, as, for example, in moving vessels when in port. The rope used for these purposes is very stout, and often a great length of it has to be drawn in. Under such circumstances a windlass would not answer the purpose, because the great quantity of rope would be inconvenient; but with the capstan, by keeping only two or three turns of the rope at a time upon the cylinder, this difficulty can be completely got over.

§ 69. **Toothed Wheels.**—It is frequently necessary to communicate rotation from one axis to a parallel axis, and there are various methods by which this can be accomplished. One of these methods has already been referred to (§ 61); another of them will be here considered.

A and B (fig. 74) are two wheels mounted on parallel axes. Each wheel is furnished with projecting teeth which enter into the corresponding spaces in the other wheel. When one of these wheels revolves, it necessarily causes the other wheel to revolve also. At the place where the two wheels are in contact it is clear that the teeth on each wheel must be moving with the same velocity. Suppose for example that 10 teeth on the wheel A pass the line

Fig. 74.

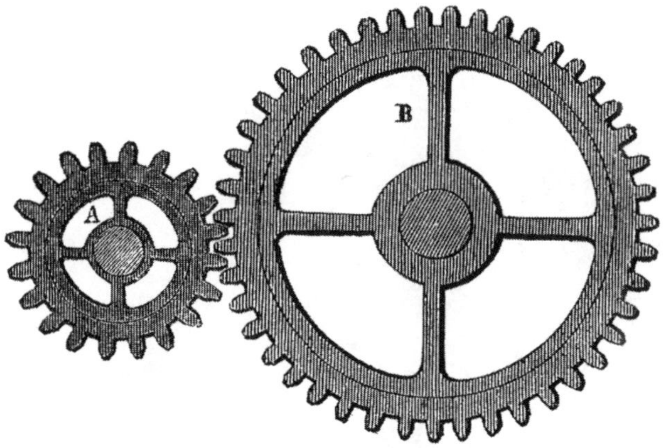

joining the centres of A and B in one minute, then 10 teeth on the wheel B will pass the same line in one minute.

Since all the teeth on the two wheels are of the same size, it appears that the number of teeth on each of the wheels will be proportional to its circumference, and since the circumferences are proportional to the radii, the number of teeth on each wheel will be proportional to its radius.

The wheel B contains 44 teeth and the wheel A 20 teeth, so that when 20 teeth have passed the line of centres then A must have made one revolution, and when 44 teeth have passed the line of centres

FIG. 75.

then B must have made one revolution. It therefore appears that when B has made one complete revolution, A must have made 44÷20 revolutions, and in general we have the following result:

When two toothed wheels are working together the

number of revolutions made by each of the wheels in a given time is inversely proportional to the number of its teeth.

§ 70. **Transmission of Force by Toothed Wheels.**—A very useful machine for lifting weights is shown in fig. 75, the principle of which can be best explained if we look only at the essential working parts (fig. 76).

FIG. 76.

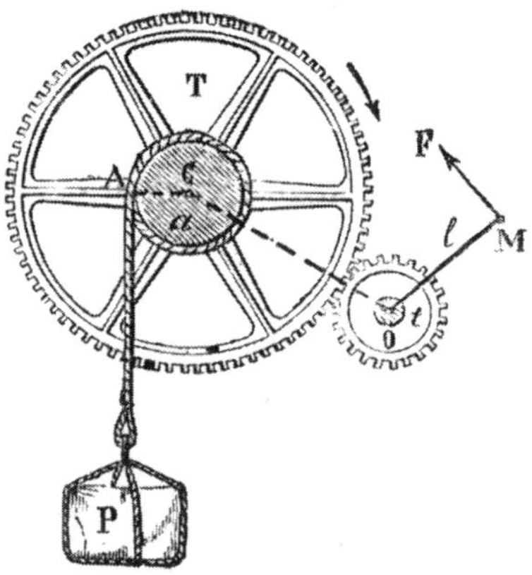

A weight P is suspended by a rope which passes at A round the cylinder *a c*. This cylinder is attached to the large toothed wheel T, into which works the smaller toothed wheel *o t*. The handle at M turns the small toothed wheel; this causes the large toothed wheel, and with it the cylinder, to revolve, and this raises the load P. We have then two distinct questions to consider :

1. Through what distance must the handle be moved to raise P one foot?

2. What power must be applied at the handle in order to raise a given load?

It will facilitate the consideration of these questions to introduce actual numbers. We shall suppose that the large wheel T has 88 teeth, and the small wheel *t* 22 teeth, and that the length O M of the handle is 12 inches, while the radius of the cylinder is 4 inches.

When the handle has made one complete revolution the wheel *t* has turned round once, but as it has only one-fourth of the number of teeth that the large wheel T has, it follows that one revolution of the handle will only produce one-fourth of a revolution in the large wheel. The weight will therefore only be raised through a distance equal to one-fourth of the circumference of the cylinder. It therefore appears that four complete revolutions of the handle will be required in order to raise the weight through a height equal to the circumference of the cylinder.

Since the radius of the circle described by the handle is three times the radius of the circle whi h represents a section of the cylinder, it is evident that the circumference of the circle described by the extremity of the handle must be three times the circumference of the cylinder. It is therefore obvious that when the load P has been raised through a height which is equal to the circumference of the cylinder, the handle M must have moved through a distance which is equal to four times the circumference of a circle which is three times the circumference of the cylinder. Therefore for each foot through which the load is raised the extremity of the handle must be pushed through a distance of 12 feet.

We now turn to the second part of the question, and inquire what power must be applied at the extremity of the handle in order to raise a given load. We may for this purpose consider that the mechanism consists of two bent levers. In the first of these levers the power is applied at M, and the load is overcome at the point of contact of the two wheels. In the second of the two levers the power is applied at the point of contact of the wheels, and the load is overcome at A. We can in the first place determine the ratio in which the point of contact of the wheels divides the line joining the centres C and O, about which the levers turn; for since one of the wheels has four times as many teeth as the other, its radius must also be four times as great as that of the other wheel. The absolute lengths of the radii are determined by the size which it is necessary to give to the wheels, in order that the teeth may be strong enough to resist the forces to which they are exposed. At present we are only concerned with the ratio of the radii, and we may suppose that the radius of the smaller wheel is 4 inches, and that of the larger 16 inches, these numbers, it will be observed, having the desired ratio.

Now by the principle of the bent lever the force applied at M perpendicularly to the arm O M bears to the load overcome the same ratio as the radius of the small wheel bears to O M, and hence it follows that the power applied at M is multiplied 3-fold when applied to the circumference of the large wheel. If, therefore, a force of 10 lbs. be applied at M along the line M F, then a force of 30 lbs. is applied to the circumference of the large wheel perpendicularly to the line of centres O C. But this force of 30 lbs. may be considered as the power applied to the longer arm of the second bent lever. And as the arms of this lever are respectively 16 inches and 4 inches, it follows that the force of 30 lbs. will be able to overcome a load at A which bears to 30 lbs. the same ratio which 16 bears to 4. The load which can be sustained at P is therefore $4 \times 30 = 120$ lbs. It thus appears that a power of 10 lbs. applied at M perpendicularly to the line O M will balance a load of 120 lbs. at P.

We may now bring together the two results at which we have arrived. We have seen, in the first place, that the power must be moved through a distance of 12 feet, in order for the load to be raised through a distance of 1 foot; and now we have ascertained that the pressure applied at M will be able to overcome a load at P which is 12 times as great.

In applying a pressure of 10 lbs. for a distance of 12 feet, a quantity of work is done which is equal to 120 foot-pounds. We therefore perform 120 foot-pounds of work against one point of the machine, and we ask how much work has been done at the other? The answer is easily found. The weight which has been raised is 120 lbs., and it has been lifted through 1 foot, and therefore 120 foot-pounds of work have been accomplished.

Here, again, we have an illustration of the fundamental truth that we cannot create energy by the use of mechanical contrivances. All we can do is to modify the form in which the work is accomplished which energy is capable of doing.

If we examined other mechanical arrangements, contrived for the purpose of augmenting our power, we should arrive at a similar result; hence we may make the following general statement:—

The ratio of the distance through which the power is exerted to the distance through which the load is raised is equal to the ratio of the load raised to the power applied.

§ 71. **The Chinese Windlass.**—An ingenious contrivance for enabling the exertion of a comparatively small power to overcome a large resistance, is presented in the machine known as the Chinese Windlass (fig. 77). One half, C, of the cylinder of an ordinary windlass is made with a somewhat smaller circular section than the other half, B. The two ends of a rope are attached to the parts B and C, and are coiled

FIG. 77.

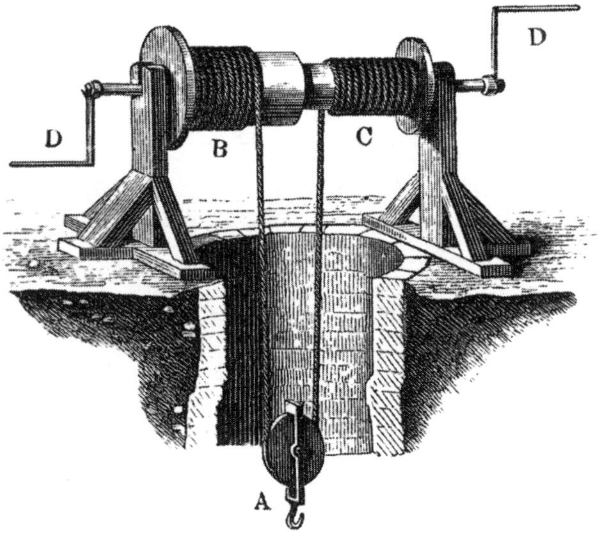

around the cylinder in the way shown in the figure. In the loop of the rope a moveable pulley A is suspended. When the handle at D is turned the rope is wound upon the cylinder B, but at the same time it is unwound from the cylinder C; owing, however, to the fact that the cylinder B is larger than C, the rope is wound up slightly faster than it is unwound, and therefore the pulley A slowly ascends.

We could ascertain the power necessary to raise a given load directly, but for the purposes of illustration we shall make use of the principle of work.

Let us suppose that the circumference of the circle described by the handle D is 5 feet, and that the effective circumferences of the cylinders B and C are respectively 2 feet and 1 foot 8 inches. When the handle has completed one revolution the power has been applied through a distance of 5 feet; consequently 2 feet of the rope have been wound up upon the cylinder B, while 1 foot 8 inches has been unwound from C. The part of the rope which is not in contact with the cylinders has therefore been shortened by the difference between 2 feet and 1 foot 8 inches, that is by 4 inches; and as this must be equally distributed between the two portions of the rope going to the pulley A, it follows that each of them will be shortened by 2 inches, and that therefore the pulley A, and consequently the load suspended from it, will be raised through 2 inches. It thus appears that when the power has been applied through a distance of 5 feet, the load will have been raised through 2 inches. If, therefore, friction be overlooked, it follows, from the principle of work, that a given power applied at the handle will be able to raise a load at A which is 30 times as great as the power.

It will thus be seen that this contrivance enables a small force to be turned into a very much larger one. The efficiency of the instrument will be the more increased according as the diameters of the two cylinders become more nearly equal. On the other hand, the machine will become more slow in its action in proportion as it becomes more powerful; in fact, it is evident that if the two cylinders were actually equal the load would never be raised at all.

§ 72. **The Differential Pulley.**—The principle embodied in the Chinese Windlass is also applied in a much more compact machine, known as the differential pulley, shown in fig. 78. In order to explain this contrivance, a diagram is shown in fig. 79, which exhibits the mode in which the machine acts.

FIG. 78.

FIG. 79.

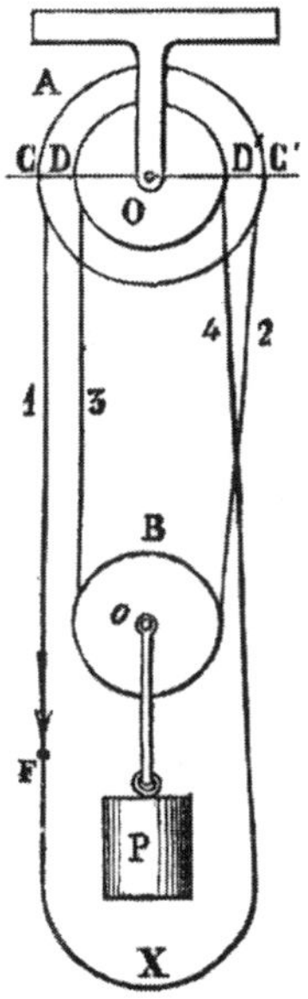

The differential pulley-block A consists of two wheels in one piece, furnished with grooves in which the chain rests. One of these grooves, C A C′, is slightly larger than the other, D D′. An endless chain passes over this block in the following way. Commencing at the point marked F, where the power is applied, the part of the chain marked 1 passes over the larger wheel in the upper block, thence down again at 2 under the lower pulley, up again at 3 over the smaller wheel, and then down at 4 to join 1.

When the chain is pulled downwards at F the upper block begins to revolve, the chain is wound in by the larger groove and lowered down by the smaller groove. As it is thus wound in somewhat faster than it is lowered, the load slowly ascends.

In order to explain the principle of this ingenious contrivance more fully, it will be well to take a special case. In a pulley-block of this kind which was employed in certain experiments the effective circumference of the large groove was 11·84 inches, and that of the smaller groove is 10·36 inches. When the upper pulley has made one revolution, the large groove must have drawn up 11·84 inches of chain, since the chain cannot slip on account of certain ridges in the groove. In the same time the small groove has lowered 10·36 inches of chain; hence when the upper pulley has revolved once the chain between the two must have been shortened by the difference between 11·84 and 10·36, that is, by 1·48; but this can only have taken place by raising the moveable pulley through half 1·48, that is, through a space 0·74. The power has therefore acted through 11·84, and has raised the resistance 0·74. The power has therefore moved through a space 16 times as great as that through which the load has moved. In fact it is very easy to verify by actual trial that 16 feet of chain must be pulled out in order to raise the load 1 foot.

If it were not for friction we should infer, from the principle of § 70, that the power need only be the sixteenth part of the load; a few trials will, however, show that the real efficiency is not so large, and that in fact *more than half the power exerted is merely ex-*

pended upon overcoming friction. This will lead to a result of considerable practical importance.

With the pulley-block at present under description it was found that a load of 200 lbs. was raised by a power of 38 lbs., that is to say, the power is about one-sixth of the load. This result was confirmed by placing different loads on the lower block; and therefore we may safely adopt the practical rule that with a pulley-block of this class a man would be able to raise a weight six times greater than he could without such assistance.

It will thus be seen that friction has an enormous effect on this description of pulley-block. Every pound of force exerted would raise 16 lbs. if there were no friction, while the presence of friction actually reduces the load raised to less than half that amount. After this statement it will seem somewhat paradoxical to assert that the differential pulley really owes a great part of its utility to the large friction present. It is a very remarkable and useful property of the differential pulley that a weight which has been hoisted by it will remain suspended without any tendency to run down. This is a source of great practical convenience, not possessed by some other pulley-blocks. The load raised by the three-sheave pulley-block, § 66, for instance, will run down unless the free end of the rope be properly secured. The difference in this respect between the three-sheave pulley-block and the differential pulley-block is not a consequence of any special mechanism; it is simply caused by the excessive friction in the latter.

§ 73. **The Screw.**—The general principle of this

very useful mechanical power may be understood from fig. 80. The screw A B is prevented from longitudinal movement, but is capable of being turned round by the handle C P. The piece D E is capable of longitudinal movement which is communicated to it by a projecting pin which enters the groove in the screw. When the screw is made to revolve, the pin is carried along and, in one complete revolution, the pin, and therefore the piece D E, will move through a distance equal to that between two consecutive threads on the screw. We hence see how the relation which subsists between the force and the load can be computed in the case of the screw by the principle of work. When the power has moved round once in the circumference of the circle which it describes, the load will have moved through a distance equal to the interval between two consecutive threads of the screw.

FIG. 80.

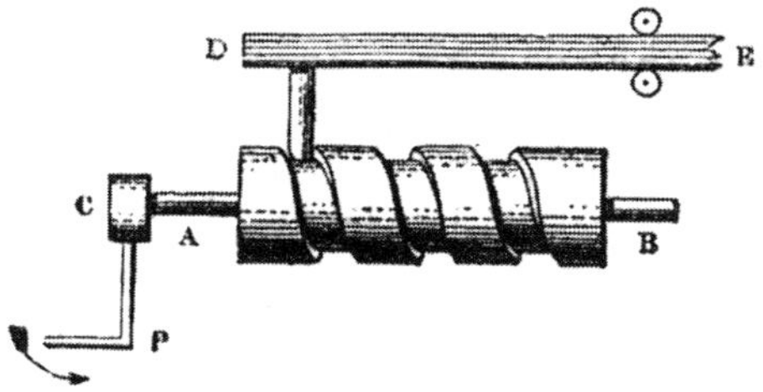

FIG. 81.

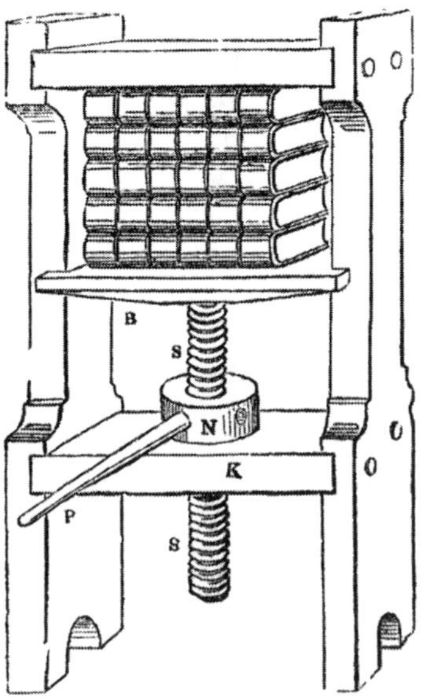

To illustrate the application of the screw for the

purpose of exerting a very great pressure through a very small space, we shall take the case of the bookbinder's press, fig. 81. The nut of the screw N is in this case turned by a handle N P. The screw itself, S S, is attached at the top to a plate, B, which can slide vertically between the two fixed supports, but the screw is not itself capable of rotation upon its axis. The books to be pressed are placed between the plate on the top of the screw and the top of the press. By turning round the handle the nut is turned round and the screw is forced upwards.

To show the great power which a machine of this description is capable of exerting, let us suppose that the interval between the threads of the screw is one third of an inch, and that the circumference of the circle which the power describes is equal to seven feet, then the distance through which the power moves is equal to 252 times the distance through which the load moves. If, therefore, there were no friction, the books would be squeezed with a power 252 times as great as that applied to the extremity of the handle. The presence of friction will probably reduce this to about one-fourth.

QUESTIONS.

1. If the handle of a windlass (fig. 71) have a radius equal to 5 times the radius of the cylinder round which the rope is wound, determine the force which must be applied to the handle to raise a load of 100 lbs. *Ans.:* 20 lbs.

2. Show that the energy expended in turning the handle is equal to the work done in raising the bucket.

3. Describe the principle of the capstan. If the length of the arms of the capstan be 12 times the radius of the cylinder, and if four men each exert a pressure of 40 lbs. upon the arms, what will be the tension produced on the rope? *Ans.:* 1920 lbs.

4. If a wheel, B, fig. 74, have 75 teeth, how many teeth should there be in the wheel A, so that it shall make three revolutions while the large wheel makes one revolution? *Ans.:* 25.

5. In a machine such as that represented in fig. 75 the radius of the circle described by the handle is three times the radius of the cylinder, and the number of teeth in the small wheel is one-seventh of that in the large wheel. If two men are working this machine, how much force must each exert to raise a load of one ton? *Ans.:* 53·33 lbs.

6. In the last case find through what distance each man must exert his force in order to raise the load through a height of 20 feet? *Ans.:* 420 feet.

7. In the Chinese windlass, fig. 77, show that the tension of the rope is one-half the load suspended from the lower block.

8. Show that the relation between the power and the load in the Chinese windlass may be deduced from the principle of the lever.

9. If the circumferences of the two wheels in the upper block of a differential pulley be 12 in. and 14 in., through how many feet must the power be exerted in order to raise the load 1 foot? *Ans.:* 14 feet.

CHAPTER XII.

THE MOTION OF A FALLING BODY.

§ 74. **Experimental Illustrations.**—We shall commence by describing certain simple experiments which can be readily made, and then we shall discuss the laws to which we are conducted by these experiments. A beginner is apt to be surprised when he is told that a heavy body and a light body will fall to the ground in the same time from the same height. Yet nothing can be easier than to prove this experimentally. Take a piece of cork in one hand, and a bullet in the other; hold your hands at the same height above the floor and drop the two objects at the same moment: they will reach the ground together. If, however, one of the objects were a feather and the other were a stone, no doubt the former object would be outstripped, and would reach the ground long after the latter. But this arises from a cause quite different from gravity: it is the resistance of the air which retards the feather. Even the stone is retarded to a certain extent, but the feather, on account of its greater extent of surface in proportion to its weight, is much more retarded. If we could get rid of the influence of the air, the stone and the feather would fall in the same time. This can actually be verified by performing the experiment (or a similar one) in a vacuum. But it can also be shown in a much more simple manner: Lay a small flat feather upon a penny piece and hold the penny horizontally with the feather on the top. If then you let the penny fall, it will be followed by the feather, which remains for the whole time in contact with the penny. In this case the penny shields the feather from the resisting influence of the air, and then it is found that the two objects fall through the same height in the same time. We are finally led to the important result that *the time occupied by a body in falling to the surface of the earth, if dropped from a point above it, is independent of the weight of the body as well as of the materials of which it is composed.*

There are, of course, certain apparent exceptions to the generality of this statement. It will not, for example, be true of a balloon, or of a live sparrow. In each of these cases the air is made to supply a force which more or less overcomes the force of gravity, but *if there were no air* then the balloon and the sparrow would each fall to the ground just as rapidly as a 56 pound weight. We shall not further refer to the resistance of the air, but shall discuss the phenomena presented by falling bodies as they occur in a space from which the air has been removed.

We have by these considerations cleared the way for an important quantitative determination. Taking a given portion of time, for example one second, we see that the distance through which a body will fall in one second depends neither upon the weight of the body nor on its composition. This distance is therefore a constant quantity for every description of body, and it is of the utmost importance to determine that quantity accurately. The value of this quantity is slightly different at different parts of the earth, though constant at each one: at any part of the United Kingdom it may be taken as 16·1 feet.

§ 75. **Velocity acquired by a Falling Body.**—It has hitherto been supposed that the stone was at rest before it was dropped, but we shall now describe an experiment in which this condition is not fulfilled. Imagine the *lift* in a hotel to be a room 16·1 feet high, then, when the lift is at rest, a stone will take 1 second to drop from the ceiling to the floor. But now suppose the experiment to be repeated when the lift is either ascending or descending with an uniform velocity. It will be found that, no matter what be the velocity of the lift (provided it remains uniform for a second), and no matter whether the lift be going up or down, the stone will still take exactly 1 second to fall from the ceiling to the floor.

To illustrate the important conclusions which can be drawn from this experiment, let us make some suppositions with reference to the velocity of the lift. Suppose that the lift is *descending* with a velocity of 5 feet per second. Then, since it is found that the stone reaches the floor in 1 second, it is manifest that during that second the stone must have fallen through the 16·1 feet which is the height of the room, and also through the 5 feet by which the floor of the room has descended in the same time. The total distance traversed by the stone is therefore $16{\cdot}1+5=21{\cdot}1$ feet. It is, however, to be observed, that at starting the stone had the same velocity as the ceiling of the lift, *i.e.* 5 feet per second, and therefore all will be explained by supposing that, in accordance with the first law of motion, the stone retained its velocity of 5 feet per second, and that meanwhile gravity acted upon the stone so as to draw it 16·1 feet nearer the earth than it would have been had gravity not acted.

On the other hand, suppose that the lift had been ascending with a velocity of 5 feet per second, then the actual distance travelled by the stone in falling from the ceiling to the floor is $16 \cdot 1 - 5 = 11 \cdot 1$ feet. This will be explained by remembering that the stone had an *upward* velocity of 5 feet per second at starting, and that therefore, in accordance with the first law of motion, it would have ascended 5 feet in 1 second had no force acted upon it. It is, then, only necessary to suppose that gravity in this case draws the body 16·1 feet nearer the earth than it would have been had gravity not acted.

We therefore see that whether the stone be at rest, or whether it be ascending or descending vertically, the effect of gravity in each case is to bring the stone in 1 second 16·1 feet nearer to the earth than it would have been had gravity not been acting.

By suitable contrivances it can be proved that a body dropped from rest will, in two seconds, move over a space of 64·4 feet. Now we have already seen that in the first second the space described is 16·1 feet. It follows that in the second second the space described by a body falling freely from rest is $64 \cdot 4 - 16 \cdot 1 = 48 \cdot 3$. It is thus obvious that the space described in the second second is three times as great as the space described in the first second. To what is this difference to be ascribed? At the commencement of the first second the body was at rest, but the body commenced its motion in the second second with all the velocity which it had acquired during the first. We are now enabled to ascertain the *velocity* of the body at the moment when the first second was completed. By the principle just explained it appears that during the second second, as during the first, the effect of gravity is simply to make the body 16·1 feet nearer the earth than it would otherwise have been. Now the body moves altogether 48·3 feet in the second second; and as the action of gravity during that second will only account for 16·1 feet, it follows that the residue of 32·2 feet must be due to the velocity acquired in the *first* second and possessed by the body at the beginning of the *second* second.

We are therefore led to the very important result that the velocity which a body acquires by gravity during a fall of one second is 32·2. It follows that if the force of gravity were entirely suspended after the body had been acted upon by it for one second, the body would then continue to move on uniformly with the velocity of 32·2 feet per second.

As a matter of fact gravity continues to act during the second second, so that the distance traversed by the body is augmented by 16·1 feet, and thus reaches the value 48·3 already specified.

It need not be a matter for surprise that, at the close of the first second the velocity acquired is 32·2, though the distance moved over during that second is only 16·1. It will be remembered that the body starts from rest, and that, as its velocity increases, the distance which it travels in a given time increases also. It therefore moves much further in the last half of the second than in the first half, and consequently the total distance travelled must be less than the distance which would have been accomplished had the body been moving during the whole second with the velocity acquired at its termination.

Let us now consider how far the body will travel in the third second. It might not be easy to ascertain this by direct experiment, but we can deduce the result by reasoning from what we have already learned. We have first to ascertain the velocity acquired by a body which has been falling for two seconds, because it is with this velocity that the third second is commenced.

Suppose that the lift already described was descending with the uniform velocity of 32·2 feet per second. If a stone be let fall from the ceiling of the room, it will, as before, reach the floor in one second. The stone will therefore have acquired, *relatively to the room*, a velocity of 32·2 feet per second. But the room is itself in motion with a velocity of 32·2 feet per second. It therefore appears that the stone, when it *began* to fall, had already a velocity of 32·2; and that during its fall it acquired an additional velocity of 32·2, so that at its close the stone actually had a velocity of $32{\cdot}2 + 32{\cdot}2 = 64{\cdot}4$ feet per second.

A body falling freely has at the close of the first second a velocity of 32·2; therefore, by the end of the second second, the body will have acquired a velocity of 64·4, and with this velocity it will commence the third second. The motion during the third second will therefore be composed of 64·4 feet due to the initial velocity, and 16·1 feet due to the effect of gravity, and therefore the total distance accomplished in the third second is $16{\cdot}1 + 64{\cdot}4 = 80{\cdot}5$.

The velocity acquired during the third second is again 32·2, which, together with the velocity 64·4 possessed at the beginning of this second, gives 96·6 as the velocity at the beginning of the fourth second, and therefore the distance traversed in the fourth second is $96·6+16·1=112·7$.

We are thus led to results which may be summed up in the following way :

Velocity at end of—

1st second	$32·2 \times 1= 32·2$
2nd „	$32·2 \times 2= 64·4$
3rd „	$32·2 \times 3= 96·6$
4th „	$32·2 \times 4=128·8$
&c.	&c.

Or in general :

A body falling freely from rest acquires a velocity which is equal to the product of 32·2 *and the number of seconds during which the motion has lasted.*

Similarly we have the following results, for the distances moved through in each of the consecutive seconds :

Distance moved over during—

1st second	$16·1 \times 1= 16·1$
2nd „	$16·1 \times 3= 48·3$
3rd „	$16·1 \times 5= 80·5$
4th „	$16·1 \times 7=112·7$
&c.	&c.

Or in general :

A body falling freely from rest moves over spaces proportional to the consecutive odd numbers (1, 3, 5, 7,) *in each of the consecutive seconds during which the motion lasts.*

We are now able to find the total space through which the body will fall in a given number of seconds. It is, in fact, only necessary to add together the spaces described during each of the consecutive experiments, and the result is exhibited in the following scheme :

Distance fallen from rest in

1 second is	$16 \cdot 1 \times 1^2 =$	$16 \cdot 1$
2 ,, ,,	$16 \cdot 1 \times 2^2 =$	$64 \cdot 4$
3 ,, ,,	$16 \cdot 1 \times 3^2 =$	$144 \cdot 9$
4 ,, ,,	$16 \cdot 1 \times 4^2 =$	$257 \cdot 6$

We are therefore led to the following very important result :

If a body be allowed to fall freely from rest, it will, in a given number of seconds, move over a distance which is found by multiplying the square of the number of seconds by 16·1.

The velocity acquired in one second by a body which has fallen from rest is usually denoted by the letter g; the value of this quantity varies slightly at different parts of the earth's surface. In London the value of g is 32·19, and it differs very little from 32·2 at any part of the United Kingdom.

§ 76. **Algebraical Expressions.**—The use of algebraical symbols will enable the results at which we have arrived to be expressed very concisely.

Let v denote the velocity acquired by a body which has been falling freely for t seconds, and let s denote the number of feet over which the body has moved; then the laws we have arrived at are thus expressed:

$$v = gt \qquad \text{i}$$
$$s = \tfrac{1}{2}gt^2 \qquad \text{ii}$$

From these equations a third can be deduced which expresses the velocity in terms of the distance.

Squaring each side of the equation i we have

$$v^2 = g^2t^2$$

and dividing this by the equation ii we deduce

$$\frac{v^2}{s} = \frac{g^2t^2}{\tfrac{1}{2}gt^2}$$

whence by obvious reductions we have

$$v^2 = 2gs \qquad \text{iii}$$

By the aid of the equations i, ii, iii, many problems relating to falling bodies can be solved.

§ 77. **Effect of the Initial Velocity.**—We have hitherto supposed that the stone or other body was simply *dropped*, and that it was not *thrown* down. If, however, the stone were thrown down, we have still the means of calculating the distance it will move over in a given time, as well as the velocity which it will acquire. In the act of throwing, the hand moves with a certain velocity, and the stone in the hand partakes of that velocity, and when released it starts off with that velocity. It will thus be observed that the effect of the act of throwing is merely to impart initial velocity to the stone.

Example.—A stone is thrown downwards with an

initial velocity of 10 feet per second, it is required to find how far it will have fallen in 1 second. From the principles already illustrated, it is clear that the total distance must be $10 + 16 \cdot 1 = 26 \cdot 1$.

The case of a stone thrown upwards appears at first sight to present somewhat greater difficulties, but this is not really the case, as all these questions can be solved by the following general principle :

A body thrown in any direction will be $16 \cdot 1\ t^2$ *feet nearer the earth, at the end of* t *seconds, than it would have been had gravity not acted upon it.*

Example.—A body is thrown vertically upwards with a velocity of 40 feet per second. Where will it be at the end of 2 seconds?

Had gravity not acted, the body would have ascended in 2 seconds to a height of $2 \times 40 = 80$

FIG. 82.

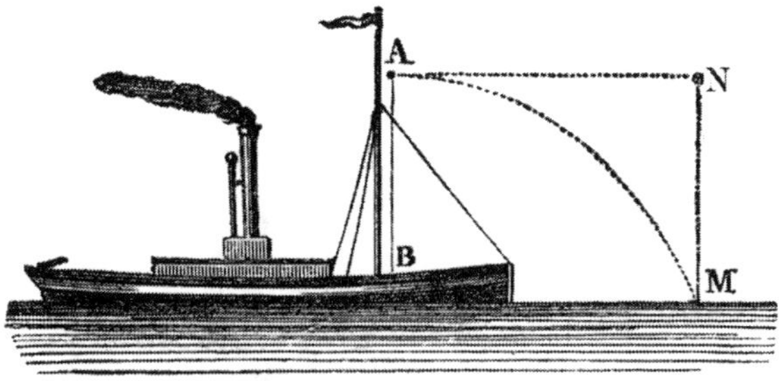

feet. The action of gravity will reduce this by $16 \cdot 1 \times 2^2 = 64 \cdot 4$ feet; hence the actual height of the body above the ground will be 15·6 feet.

§ 78. **Motion not in a Vertical Line.**—We have hitherto referred only to motion in a vertical line; it will now be necessary to examine some cases in which the motion of the body is not so restricted. From a point, A, on the mast of a steamer (fig. 82), a ball is dropped, and it falls upon the deck at a certain point, B. We may for simplicity suppose that A is 16·1 feet above the deck. When the steamer is at rest, the time taken by the ball to fall will of course be one second, and its path, A B, will be vertical. But when the steamer is in motion, with a uniform velocity, it is found that the ball still falls precisely on the same spot of the deck as when the steamer is at rest, and that the time occupied in the descent is exactly one second. If we suppose that in one second the motion of the steamer is such as to carry the mast from A B to N M, then the actual path described by the ball in its descent must be a curved line such as A M. We therefore see that when the steamer is in motion the ball moves along the curved path A M in the same time as it would take to fall vertically through the height A B. If gravity had not acted upon the ball, the latter would in one second have arrived at N, but when gravity does act the ball reaches M. Now the distance N M is obviously equal to A B, and is therefore 16·1 feet; and hence we see that gravity has brought the ball 16·1 feet nearer the earth in one second than it would have been had gravity not acted. The same result would be found whatever was the height of the mast, and therefore we are led to the general conclusion that—

A body projected horizontally will at the end of t *seconds have fallen through a space of* $16{\cdot}1\ t^2$ *feet.*

An experiment illustrating this principle may be made in an exceedingly simple manner. Take a marble in each hand, and throw one of the marbles horizontally at the same moment as you drop the other. If your two hands be at the same height, then you will find that the two marbles reach the ground together. Suppose, for simplicity, that the height of your hands above the floor is 4 feet; then the time taken by one of the marbles in falling is half a second. But, as both marbles reach the ground together, the experiment shows us that a body at a height of 4 feet from the ground will, if projected *horizontally*, reach the ground in half a second. Observe, also, that this is true whatever be the magnitude of the velocity, *i.e.*, whether it be 5, 10, or any number of feet per second.

§ 79. **Motion of a Projectile.**—We have now studied experimentally the effect of gravity upon a body which has been projected either in a *vertical* line or in a *horizontal* line. We have found that in each case the effect of gravity is to bring the body $16{\cdot}1\,t^2$ feet nearer the earth in t seconds than it would otherwise have been. We are therefore tempted to inquire whether the same statement would not be true for a body projected *in any direction*. In every way in which we can test this suggestion we find it to be verified, and there cannot therefore be the slightest doubt that it is true.

To illustrate the application of this principle, we shall consider the case of a ball projected in any direction, and we shall show how the curved path in which the ball will move can be deduced by calculation.

Let O (fig. 83) be the point from which the ball is projected in the direction O P_1. Then, in consequence of the first law of motion, we should find that if it had not been for gravity the ball would reach P_1 in one second, P_2 in two seconds, P_3 in three seconds, and so on, where $O P_1 = P_1 P_2 = P_2 P_3$, &c. But if we wish to find where the body will actually be at the end of t seconds, the principle just laid down will solve the question. Take, for example, $t = 3$. If gravity had not been acting, the body would in three seconds

FIG. 83

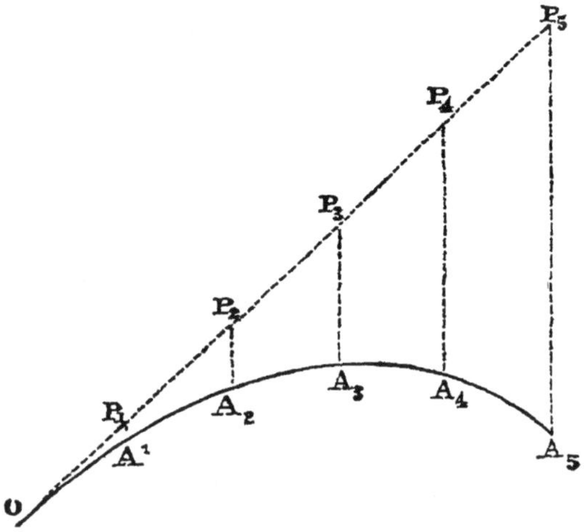

have reached the point P_3. We can find where the body actually is by taking a point A_3, which is vertically beneath P_3 at the distance $16{\cdot}1 \times 3^2$ feet. Similarly we can find where the body is after any other specified number of seconds, and thus we obtain the points A_1, A_2, &c. It is therefore obvious that the actual path which the body describes is represented by the curved line O, A_1, A_2, A_3, A_4, &c. This curve is known to mathematicians as a *parabola.*

QUESTIONS.

1. If a ball of iron and a ball of wood be dropped at the same moment from the same height, will they reach the ground together?

2. Though a penny piece and a feather give a different result, show that this does not invalidate the general law, that, so far as gravity is concerned, the time taken by a body in falling is independent of the mass and material of the body.

3. If the velocity acquired by a body in falling freely for one second be g, what will be the velocity at the end of 2 seconds or 3 seconds? *Ans.:* $2g$ or $3g$.

4. Give the steps by which it is shown that a body falls from rest through $16 \cdot 1t^2$ feet in t seconds.

5. Show that the number of feet through which a body falls from rest is very nearly equal to the square of the number of quarter seconds occupied in the descent.

6. Find the velocity acquired by a body which has fallen from rest through a height of h feet. *Ans.:* $\sqrt{64 \cdot 4h}$.

7. A body is projected downwards with a velocity of 40 feet per second. How far will it have descended in 3 seconds? *Ans.:* 264·9 feet.

8. A body is projected vertically upwards with a velocity of 96·6 feet per second. How long will it be before it reaches the ground? *Ans.:* 6 seconds.

9. How high will the body referred to in the last question ascend? *Ans:* 144·9 feet.

10. What will be the velocity of the body after it has been in motion for 5 seconds? *Ans.:* 64·4.

11. A body is projected horizontally from a height of 64·4 feet with a velocity of 20 feet per second. After what time will it reach the ground, and at what distance is the point where it falls on the ground from the point vertically beneath the point of projection? *Ans.:* 2 seconds; 40 feet.

CHAPTER XIII.

THE SECOND LAW OF MOTION.

§ 80. **Mass.**—It will be necessary, before discussing the second law of motion, to explain clearly what is meant by the word *mass*. This word is sometimes regarded as a synonym with *weight*, but the latter term is open to an ambiguity which, though not productive of inconvenience in ordinary language, is apt to be a source of confusion to the student of mechanics. The word *weight* is sometimes used to denote a certain quantity of *matter*, and sometimes to denote a certain *force*. In the former case the ambiguous word *weight* may be replaced by the word *mass*. In the latter case we may use the word *gravitation*.

An illustration will show the necessity for this distinction. The *mass* of a piece of iron is, of course, a constant quantity at whatever part of the earth the iron be placed; but the *gravitation* of the piece of iron depends not alone upon the iron, but also upon the earth by which the attraction is produced. The gravitation of the iron will, indeed, be different at different places on the earth.

As a matter of fact, we are very much in the habit of measuring the *mass* of a body by its *gravitation*, and hence these two ideas have become somewhat confused. In the use of a common pair of scales we make use of the principle that *at the same locality equal masses have equal gravitations*.

§ 81. **Action of Force upon Mass.**—A *mass* of

one pound is allowed to fall from rest. The force which acts upon that mass is the *gravitation* of one pound. We thus see that when a force equal to the gravitation of one pound acts upon a mass of one pound for one second a velocity g is produced.

We have, however, now to consider a question of quite a different kind. Suppose that a mass of 10 lbs. be acted upon by a force equal to the gravitation of 1 lb., what velocity will be acquired in one second? If the force were equal to the gravitation of 10 lbs., then, of course, the velocity produced would be g. The question for us to consider is the velocity which a force only one-tenth of this amount will be able to produce.

§ 82. **Experiments with an Inclined Plane.**— Construct an inclined plane, of which the length is 10 times the height. Let this plane be carefully polished, and procure a small truck mounted on nicely-fitted wheels so that the friction is inappreciable. Let the truck be laden so that the mass of the truck and its contents taken together shall be equal to 10 lbs. Then, when the truck is placed upon the inclined plane so that it would run straight down if permitted to do so, a certain force must be expended to keep the truck from moving. It appears from the principle of the inclined plane (§ 54), that the force which must be exerted parallel to the plane in order to keep the truck up, must bear to the gravitation of the truck the same proportion which the height of the plane bears to its length. The force, therefore, which must be exerted parallel to the plane is equal to 1 lb. We thus see that in the motion of the truck down the plane a mass of 10 lbs. is urged by a force of 1 lb., and therefore the observed motion of the truck will give us the information which we desire.

When suitable arrangements for measuring the motion of the truck are introduced, it is possible to ascertain with considerable accuracy the velocity which is acquired in one, two, or three seconds. The results are as follows :—

In 1 second		$g \div 10$
„ 2 seconds		$2g \div 10$
„ 3	„	$3g \div 10$

We are led to a very important result when we compare these observations with what we found in the case of a body falling freely. In the latter case a force of 10 lbs. acting on a mass of 10 lbs., imparts a velocity g in one second, while in the former a force of 1 lb. acting on a mass of 10 lbs. imparts a velocity equal to $g \div 10$.

By varying the inclination of the plane it will always be found that the velocity acquired in a single second is proportional to the height of the plane, and as the effective force of gravitation down the plane is also proportional to the height of the plane, we have the following important result.

A force applied to a body at rest will in one second impart to that body a velocity which is directly proportional to the force.

By varying the mass of the truck and repeating the experiments, we obtain a series of results which may all be included in the following law.

A force applied to a body at rest will in one second impart to that body a velocity which is inversely proportional to the mass.

§ 83. **Atwood's Machine.**—These principles are of such great importance, that it will be well to point out means by which they can be verified in a different manner. For this purpose a contrivance may be used which is known as Atwood's Machine. Freed from the complication of details, the essential features of this instrument are extremely simple. A pulley, D C (fig. 84), is mounted with great delicacy, so that the friction with which its movements are retarded is practically insensible. On the rim of this pulley is a groove which carries a very flexible silken cord, to which the weights A and B are fastened. Attached to the vertical supports, are graduated scales by which the distance moved through by each weight in a given number of seconds can be ascertained. When the weights A and B are equal, they of course remain at rest. If, however, any addition be made to either weight, this one will descend and the other will rise.

Fig. 84.

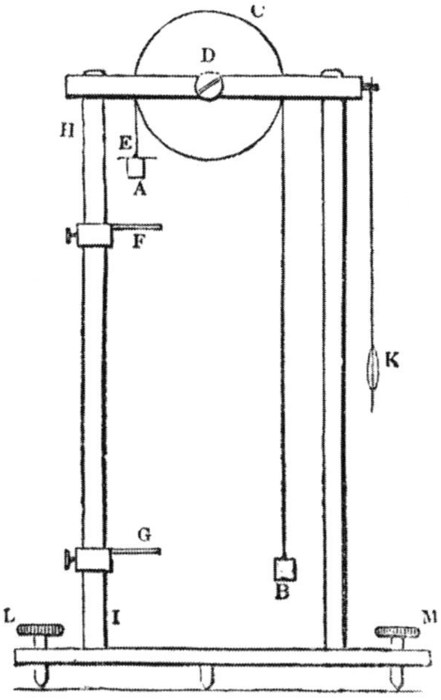

Let A and B be each 4·5 ounces, and let one of

the weights, A, receive the additional piece shown at E, which we shall suppose to weigh 1 ounce, A will therefore descend and, of course, lift up B. In order to apply the results given by Atwood's machine to other cases we are obliged to make an assumption which experience has shown to be true (§ 94). If a force equal to the gravitation of E were applied to a single mass which was equal to the masses of A, B and E taken together, a certain velocity would be produced in one second. The assumption now made is that this velocity is really identical with that which is produced in one second in Atwood's machine. With this assumption we may apply the movements observed in Atwood's machine to ascertain the motion which would be produced in a mass equal to the sum of the masses A, B and E by a force equal to the gravitation of E. In the present case the total weights would amount to 10, and the effective force is 1; we may therefore apply the indications of the machine to give us information about the motion of a mass of 10 ounces when acted upon by a force equal to the gravitation of 1 ounce. The distance moved in one second is seen to be $g \div 20$. If we vary the mass of E, though still retaining the sum of the masses constant, we can verify the law that *the distance moved over in a given time is directly proportional to the force.*

A different mode of making the experiments is also instructive. If A and B are each equal to 9·5 ounces while E is still 1 ounce, the total mass set in motion is 20 ounces. In this case it is found that in one second a distance of $g \div 40$ is accomplished. As the force in this case is the same as before, it appears that when the mass is doubled the distance accomplished in a second is halved. Making other changes in the masses, and observing the distances moved through, we infer generally that, *if the force remain constant, the distance described in a given time is inversely proportional to the mass.*

By the use of Atwood's machine, we are enabled not only to find the distance which is accomplished in a given time, but also the velocity which has been acquired during that time. For this purpose certain arrangements are made, which are also shown in the figure. As the weight A is descending in consequence of its having received the additional load E, it passes through a ring placed at F. The ring detains the piece E, so that when A has passed through the ring it is no longer acted upon by any force, because its gravitation is counterbalanced by the tension on the cord produced by the equal weight B. Under these circumstances the weight A, in accordance with the first law of motion, continues to move on uniformly with the velocity which it had at the instant when the piece E was removed by the ring. To ascertain the amount of this velocity, a small stage G is attached to the vertical rod I H, which is capable like the ring of being adjusted at any height which may be desired. By counting the oscillations of the pendulum K, the time can be determined which intervenes between the moment when A passes through the ring and its arrival at the stage. As the velocity of A is uniform between the ring and the stage, and as the distance is known by means of the graduated scale on the rod I H, we are enabled to determine the velocity with which the weight A passes through the ring.

There are three problems which may be solved by this contrivance, and we shall consider them separately.

First.—To determine the connection between the velocity which has been acquired and the time during which the force has been in action.

By placing the ring in proper positions, it can be arranged that the extra weight E shall be lifted off from A after one, two, three, &c. seconds from the commencement of the movement. In this way it is found when the weights A, B and E are unchanged, that *the velocity which is acquired is proportional to the time during which the force has been in action.*

Second.—To determine the connection between the velocity which has been acquired and the magnitude of the force by which the velocity has been produced.

For the sake of illustration, suppose that the masses of A and B are each represented by 10, while that of the piece E is 1, and suppose that the movement has been going on for one second before E is removed by the ring. The velocity acquired is then found to be $g \div 21$ feet per second. If the piece E had a mass represented by 2, while those of A and B are each 9·5, the total mass set in motion is 21 as before, but the force is now 2, and the velocity acquired in one second is double what it was before, or $2g \div 21$ feet per second. In the same way, if the mass of E were 3, while the masses of A and B were each 9, it would be found that the velocity acquired in one second would be threefold what it was in the first instance. We therefore see that *the velocity acquired in a given time by a force acting on a given mass is directly proportional to the force.*

Third.—To determine the connection between the velocity which has been acquired in a given time and the mass on which a given force acts.

Let the masses of A and B be each 10 and that of E be 1, then the velocity acquired in one second is $g \div 21$. If E be retained at 1 while A and B are each reduced to 5·25, the total mass is reduced to half of what it was at first. The velocity acquired in one second is then found to be $2g \div 21$, or double what it was before. We thus find that the *velocity produced by a given force acting for a given time is inversely proportional to the mass.*

§ 84. **Algebraical Expressions.**—The general results of experiments such as those just described may be concisely stated in the form of an algebraical expression, for the velocity v which is acquired by a mass m when acted upon by a constant force f for a time t. We shall suppose that t is expressed in seconds v in feet per second, and m in pounds. With reference to the units in which f is expressed a caution is necessary: f is a force, and must be expressed in terms of a certain unit of force. The unit of force is the *gravitation* of a mass of one pound. It is necessary to understand clearly that the *unit of force* is a magnitude of a totally different description from the *unit of mass,* in terms of which m is measured. As each of these units is indifferently termed a pound, it is the more necessary to use great caution.

The experiments have shown us that the velocity acquired is directly proportional to the force and time and inversely proportional to the mass. We therefore have

$$v = xft \div m$$

where x is a numerical quantity, which must be determined. We can employ for the determination of x the particular details of any of the experiments which have been described. The most convenient is that in which a body of mass 1 lb. is allowed to fall freely from rest for one second. In this case the velocity acquired is g, and as f, t, and m are each unity, we must have

$$g = x,$$

and therefore finally

$$v = gft \div m.$$

§ 85. **Expression for the Distance.**—It remains to determine the expression for the distance which the body will move over in a given time. We shall suppose as before that the body is initially at rest. On a line O A set off a distance O A (fig. 85), which shall express the time t according to some unit of length which may be conveniently chosen. Suppose, for example, the number of inches in O A is taken equal to the number of seconds in t. At A draw a line A B perpendicular to O A, and take the point B so that the length A B shall denote the final velocity v. Suppose, for example, that the number of inches in A B is equal to the

Fig. 85.

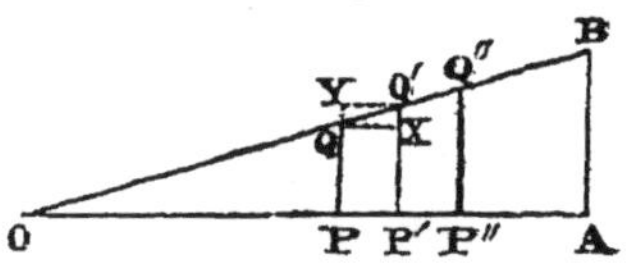

number of feet per second in the final velocity. Join O B, then the velocity of the body at any given moment of time can be immediately found. For take the point P so that the length O P represents the given time, and draw P Q perpendicular to O A. Then from the similar triangles we have

$$QP : BA :: OP : OA$$

but we have already seen that the velocity acquired is proportional to the time, wherefore the velocity at the time corresponding to P must bear to the velocity at the time corresponding to A the ratio O P : O A, which is equal to the ratio Q P : B A; but we have agreed that the velocity at A shall be represented by A B, and therefore the velocity at P will be represented by Q P. If we take any other point P′, the velocity at the time which corresponds to P′ is represented by the line P′ Q′. The distance described in a given interval of time would, if the velocity were uniform, be equal to the product of the velocity into the time. At the time P the body has the velocity denoted by P Q. If it were to retain this velocity for the time represented by P P′, the distance accomplished would be represented by P Q × P P′, that is, by the area of the rectangle P Q X P′. On the other hand, if the body had the velocity P′ Q′ for the time P P′, the distance traversed would be P′ Q′ × P P′, or the area of the rectangle P P′ Q′ Y. As a matter of fact the velocity during the time P P′ is not constant, but increases from P Q at the beginning to P′ Q′ at the end. It follows that the actual distance which the body moves over in the interval of time P P′ must be less than the number of units in the area P P′ Q′ Y, and greater than those in P P′ X Q. We therefore have a higher and lower limit to the distance accomplished in the time P P′. If P′ be taken excessively close to P, the two limits are exceedingly close together, but however close they may be, the area of the four-sided figure P Q Q′ P′ is always intermediate to them. We may therefore assume that when P and P′ are indefinitely close together, the space described in the time P P′ is represented by the area of the figure P Q Q′ P′. It can similarly be shown that during the next small interval of time P′ P″, the distance described is represented by the area of the trapezium P′ Q′ Q″ P″, hence the distance described during the time P P″ is represented by the area P Q Q″ P″. By similar reasoning the distance described in the whole time O A is shown to be represented by the area O A B. But since the triangle O A B is half of the rectangle whose base is O A and whose height is A B, we see that

$$OAB = \tfrac{1}{2} OA \times BA$$

or, putting s for the space represented by the area O A B,

$$s = gft^2 \div 2m$$

whence we derive the following important result :

A body initially at rest, whose mass is m *pounds, being acted upon by an uniform force of* f *pounds for a time of* t *seconds, will at the end of that time have moved over a distance* $gft^2 \div 2m$.

§ 86. **Effect of Initial Velocity.**—We have hitherto supposed that the body was initially at rest. We must now see how the above expressions must be modified when this condition is not fulfilled. Let the body have a velocity v' at the commencement of its motion, and let the direction of this motion be the same as the direction in which the force is tending to move the body, it is proposed to find the velocity which the body will have acquired at the end of the time t, and also the distance over which it will have moved. Observation has shown that the force f acting upon a mass m, will in the time t give that body a certain *increase* of velocity. *The amount of that increase is found to be totally independent of any velocity which the body may have previously possessed, and is equal to* gft ÷ m. It therefore appears that if the body had initially the velocity v', its velocity v at the time t must be found by the formula,

$$v = v' + gft \div m.$$

The distance which is accomplished in the time t is equal to the distance $v't$, which would have been accomplished if the body had possessed the initial velocity alone, augmented by the distance $gft^2 \div 2m$ which would have been accomplished in consequence of the force. The total distance s is therefore given by the formula

$$s = v't + gft^2 \div 2m.$$

If the initial velocity had been in the opposite direction to the force, the formulæ would have been

$$v = v' - gft^2 \div m$$
$$s = v't - gft^2 \div 2m.$$

The various results to which we have been conducted in this chapter lead to the enunciation of a very important truth, which is known as the Second

Law of Motion. This law, though capable of being concisely stated, really embraces some of the most fundamental truths in nature. Special attention is requisite, not only on account of the great importance of the subject, but also because of the difficulty of thoroughly understanding a proposition of such wide generality.

§ 87. **Momentum.**—The product of the mass of a body and the velocity with which that body is moving has received the special name of *momentum*. Thus taking the units we have adopted it appears that the momentum of a mass of m pounds moving with a velocity of v feet per second is represented by the expression mv. The experiments with Atwood's machine have led to the formula

$$v = gft \div m$$

whence $$mv = gft$$

The *momentum* acquired by the mass when acted upon by the force f for the time t is therefore gft. This result will provide us with a simple method of comparing the efficiency of two forces f and f'. Let M and M' denote the momenta which would be generated by the forces f and f' respectively if each were to act for the same time t. Then we have

$$M = gft$$
$$M' = gf't$$

whence $$M : M' :: f : f'.$$

We therefore see that the momentum produced by a force acting for a given time is proportional to the magnitude of the force. It would thus appear that the most natural method of measuring a force would be founded upon ascertaining the momentum which that force would be capable of generating in a given time.

A body may have been moving before the force is applied, it is therefore necessary to consider in what way the initial momentum of the body is affected by the subsequent application of the force. This is simple enough in cases where the direction of the initial velocity is the same as that in which the force acts. Let v' be the initial velocity, then we have as before

$$v = v' + gft \div m$$

and multiplying by m :

$$mv = mv' + gft.$$

The original momentum was mv', and the change in momentum is gft, whence we may assert that the *change in momentum is proportional to the force.*

§ 88. **Change in Momentum.**—The cases where the direction of the force does not coincide with the direction of the original velocity are somewhat more difficult. It is still true that the *change in momentum* is proportional to the force, but it is necessary to understand very clearly what is really meant by the change in momentum. Let the line O A represent both in magnitude and direction the original momentum of the body (fig. 86). After the lapse of a stated interval of time, for example one second, suppose that the force has so modified the original motion that the momentum with which the body is actually moving is represented both in

Fig. 86.

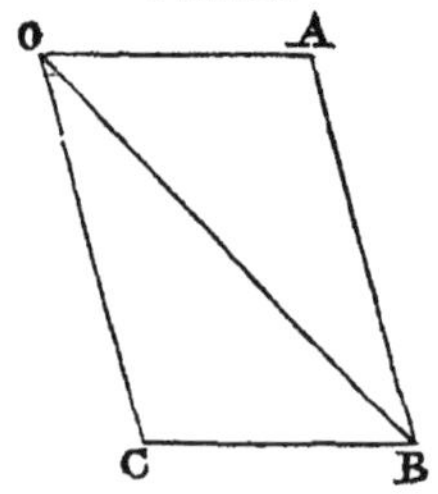

magnitude and direction by the line O B. Let O C and B C be drawn parallel to B A and O A respectively. Then the momentum represented by the line O C is defined to be the *change of momentum.*

§ 89. **Second Law of Motion.**—It is invariably found by direct experiment when practicable, or more usually by indirect methods of combined observation and calculation, that the change of momentum as above defined lies in the direction of the force and is proportional thereto. Thus in the case just considered, where the velocity which was originally represented by O A is after one second represented by O B, the force, if constant in magnitude and direction, must have been parallel and proportional to O C. Uniting all the different cases together, we have the following general statement which constitutes what is called the Second Law of Motion:—

Change of momentum is proportional to the impressed force, and takes place in the direction of the straight line in which the force acts.

Like the first law of motion, the second law really only expresses an empirical truth. It is futile to attempt to give any demonstration of these laws except what directly or indirectly has been afforded by our experience. But experience has been so unvarying in its confirmation of the truth of these laws, that no hesitation whatever can be felt in making use of them. They may in fact, as far as our universe is concerned, be regarded as axioms, which though not self-evident, are sufficiently established by experience.

One difficulty which the student may feel in making use of the second law of motion is connected with the change of momentum. He may imagine that he ought to find some *proof* that the change of momentum is represented by O C (fig. 86). This however is a mistake; *we define the change of momentum to be the momentum represented by the line* O C, and what our experience shows us is that the change of momentum *so defined* fulfils the second law of motion.

§ 90. **Illustrations.**—Suppose a body whose mass is 3 lbs. to be moving with a velocity of 5 feet per second. Let an uniform force act upon the body in the direction of its motion for one second. If the velocity of the body be now 20 feet per second, what must be the magnitude of the force?

The original momentum of the body is $3 \times 5 = 15$, and the momentum after the force has acted for a second is $3 \times 20 = 60$. The change in momentum is $60 - 15 = 45$.

The second law of motion tells us that the force is proportional to the change of momentum, and therefore the required force is 45 times the force which would generate the unit of momentum in the unit of time. To evaluate this force in pounds we know that a mass of 1 lb. will in falling from rest acquire in one second a velocity g. A force of 1 lb. will therefore produce g units of momentum in one second, and therefore the required force is $45 \div g$.

A body whose mass is 25 lbs. is projected horizontally with a velocity of 10 feet per second. If the action of gravity be considered, determine the velocity of the body and the direction in which it is moving at the end of one second.

The change of momentum being proportional to

the force and in the direction thereof, must in this case be 25 g units of momentum directed vertically downwards. This may be represented by O B (fig 87), where O A is a horizontal line proportional to the initial momentum of 250. Remembering the definition of change of momentum, it is obvious that the total momentum at the end of one second is represented both in magnitude and direction by O P. We have obviously $OP^2 = OA^2 + OB^2$, whence taking g at 32·2

FIG. 87.

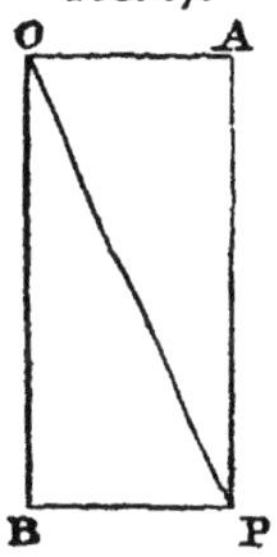

$$OP^2 = (250)^2 + (805)^2$$
$$OP = 843$$

The momentum of the body will therefore be augmented to 843, and the velocity will be $843 \div 25 = 33{\cdot}72$. The direction of the motion at the end of one second may be found by measuring the angle A O P. It is equal to 72° 45′.

In estimating the change of momentum produced by a given force, account must, of course, be taken of the time during which the force has been in action. The change of momentum is proportional to the time as well as to the force, and is therefore proportional to their product. Any ambiguity may be cleared away by making use of the following formula:

Let M be the change in momentum expressed in feet per second, f the force in pounds, and t the time in seconds, while g denotes, as before, the velocity acquired by a body in falling freely from rest for one second.

Since M is proportional to the product of f and t, we must have

$$M = xft,$$

but when $t = 1$ and $f = 1$ lb. we must have M equal to g, whence $x = g$, and finally

$$M = gft,$$

or by a simple transformation

$$f = M \div gt.$$

By this equation we are enabled to determine the force by the action of which during a given time a given change of momentum has been produced. As the effect of a given force in generating momentum is quite independent of gravity, it may seem strange to find the constant g contained in the formula. This arises from the circumstance that we have chosen the *gravitation* of a certain weight (*i.e.* 1 lb.) as our unit of force, and therefore the expression of any force in terms of this unit must involve gravity. We have thus a very convenient method of calculating the change of momentum when the force and the time are known, or the force when the time and the change of momentum are known. This may be illustrated by the following problem:

A body which weighs 4 lbs. and had an initial velocity of 10 feet per second is found after 3 seconds to be moving with a velocity of 50 feet per second in a direction which is inclined at an angle of 30° to the original direction. Determine the direction and magnitude of the constant force by which this change was produced.

Draw O A (fig. 88) to represent the original momentum, and O B to represent the final momentum, the angle A O B being equal to 30°. Then completing the parallelogram, we find that the change of momentum, according to our definition of that phrase, is represented by O P. The second law of motion asserts that the direction of the force must have been parallel to O P, and the magnitude of the force proportional to O P, and we have generally—

FIG. 88.

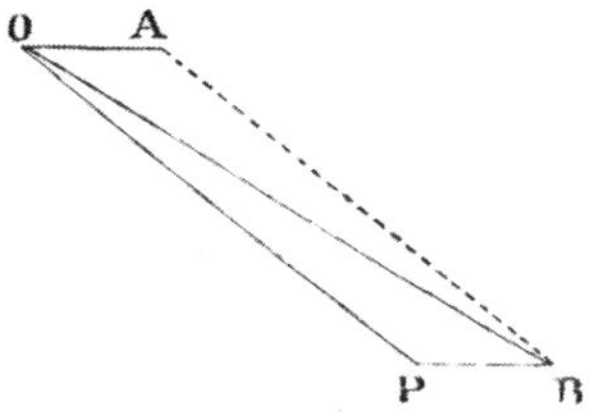

$$M = gft;$$

whence in the present case—

$$f = \text{O P} \div 96{\cdot}6.$$

If the figure be constructed, the length O P can be measured and will be found to be 40·4, whence

$$f = 0{\cdot}42 \text{ lbs.}$$

§ 91. **Motion in a Circle.**—Another very important illustration of the principles we have been considering is presented in the study of the motion of a body in a circle.

It is asserted in the first law of motion that a body once moving uniformly in a straight line will continue to move on uniformly in the same straight line unless it be acted upon by force. We may, therefore, infer, that if the motion of the body does not remain uniform some force must be in action, and we have already explained how the magnitude of the force is to be ascertained by measurement of the change in velocity. But there is another way in which a departure from the most simple form of movement may be manifested. Instead of moving in a *straight* line the body may be moving along a *curved* line, and when this is found to be the case we can with certainty conclude, in accordance with the first law of motion, that *some force* must have been in action. The second law of motion will afford us the means of ascertaining *what the force must be* when we know the shape of the curve in which the body is moving as well as the fluctuations in its velocity.

It will, however, be sufficient for our purpose to consider the question under the most simple form which it is capable of assuming. We shall suppose that the body is moving in a *circle*, and that the velocity with which it is animated is *uniform*, and what we have to ascertain is the force which acts upon the body so that its path shall remain circular and its velocity remain uniform. The first point to be demonstrated is that the force is *constantly directed towards the centre of the circle.* This is indeed obvious from the symmetry of the case, for there is no reason why the direction of the force should pass at one side of the centre rather than at the other. It is, however, easy to demonstrate the same property as a

FIG. 89.

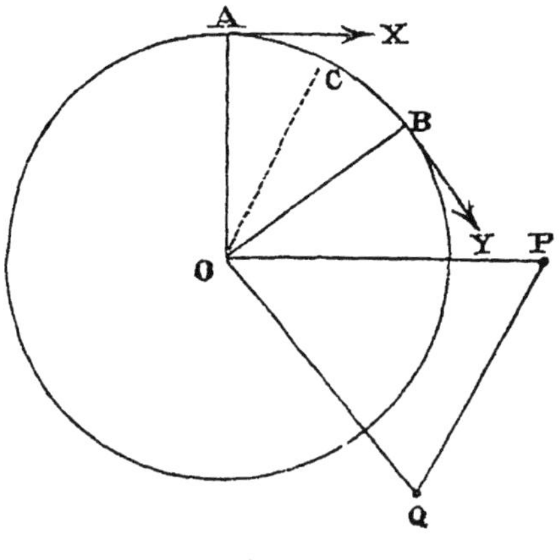

direct consequence of the second law of motion. Let A, B be two positions of the body (fig. 89), and let A X and B Y be the tangents to the circle at A and B. Then the body at A may be momentarily considered to be moving in the direction of the tangent A X, and at B in the tangent B Y. Draw through O the equal lines O P and O Q, parallel to A X and B Y, and proportional to the momentum of the body. Join P Q: then the change of momentum is represented, both in magnitude and direction, by P Q. But if O C be the bisector of the angle A O B, then the line P Q is easily seen to be parallel to O C. The change in the momentum of the body between A and B is therefore parallel to the bisector of the angle A O B. If the change in momentum between A and B had been produced by a force which was constant in direction and magnitude, that force must have been parallel to O C. Let us now suppose that A and B are so exceedingly close together that the lines O A, O B, O C are indistinguishable, then it is obvious that the change of momentum, and consequently the acting force, must lie in the direction of the radius.

§ 92. **Magnitude of the Force.**—It remains to determine the magnitude of the force in terms of m the mass of the body, v the velocity with which it is moving, and r the radius of the circle. We shall suppose that A and B are exceedingly close together, and that the distance A B is traversed in the short time t. Since O P and O Q are perpendicular to O A and O B respectively, the angle P O Q is equal to the angle A O B, and therefore taking the arc A B so small that it can be regarded as identical with its chord, we have from the similar triangles O A B and P O Q :

$$PQ : OQ :: AB : OA$$

Substituting for O Q, A B, and O A the values $m v$, $v t$, and r respectively, we deduce :

$$PQ = m v^2 t \div r,$$

but P Q is the change of momentum produced by the force in the time t, hence from § 90 we have for the force f the expression :

$$f = PQ \div g t$$

or by substitution :

$$f = m v^2 \div g r.$$

The result at which we have arrived may be thus stated :

If a body whose mass is m pounds be revolving in a circle of radius r feet, with an uniform velocity of v feet per second, the body must be constantly urged by a force which is directed towards the centre of the circle, and when expressed in pounds is equal to $m v^2 \div g r$.

§ 93. **Example.**—If a body weighing 10 lbs. be moving in a circle whose radius is 8 feet, with a velocity of 9 feet per second, what force must be constantly exerted towards the centre of the circle? In this case $m = 10$; $v = 9$; $r = 8$; and we may take g to be 32·2, whence the answer is $10 \times 9^2 \div 8 \times 32{\cdot}2 =$ 3·4 lbs.

Suppose that we had a smooth horizontal plane on which a weight of 10 lbs. could slide without friction. If this weight were fastened to one end of a string 8 feet long, while the other end of the string was attached to a point in the plane, and if the weight were then started off to revolve in a circle with the velocity of 9 feet per second, the motion would continue indefinitely, and the tension of the string (which we can easily conceive to be shown by a dynamometer) would be 3·4 lbs. In this case the tension of the string will constitute the force which, acting upon the weight, causes it to move in the circle.

If the velocity of the weight were doubled, the other data remaining the same as before, the tension of the string would be quadrupled.

We might also imagine the movement to take place in a different manner: Construct a small circular railway of 8 feet radius on a horizontal plane, place on it a truck which weighs 10 lbs., and start off this truck by an impetus which will give it the initial velocity of 9 feet per second, then as the railway is supposed to be frictionless, the truck will continue moving round and round uniformly without receiving any other impulse. In this case the reaction of the rails against the wheels of the truck will provide the necessary force which is directed towards the centre of the circle.

Questions.

1. If a truck run down an inclined plane whose height is the nth part of its length, what velocity will be acquired in one second if friction be neglected? *Ans.:* $g \div n$.

2. What distance will be accomplished by the truck in t seconds. *Ans.:* $\frac{1}{2} g t^2 \div n$.

3. Show that the velocity the truck will have acquired at the moment when it reaches the bottom of the plane is equal to the velocity which it would have acquired by falling freely from rest through a distance equal to the height of the plane.

4. Determine the velocity which a force of 1 lb. will impart to a mass of 20 lbs. in 3 seconds. *Ans.:* $3g \div 20$.

5. A force which in one second can generate a certain velocity in a certain mass will in the same time only generate half the velocity in another mass. Compare the two masses. *Ans.:* 1 : 2.

6. If weights of 6 ounces and 7 ounces be attached to the extremities of a cord which passes over a frictionless pulley, how far will the heavier weight descend in t seconds? *Ans.:* $\frac{1}{2}gt^2 \div 13$.

7. What velocity will be acquired in the same time? *Ans.:* $gt \div 13$.

8. If the two weights be each 4 ounces, and if an additional weight of one-tenth of an ounce be added to one side, in what time will a distance of 4 feet be accomplished? *Ans.:* 4·5 secs.

9. The sum of the two weights in Atwood's machine is 10 ounces and the velocity acquired in 3 seconds is 2 feet per second, what must the weights be? *Ans.:* 5·104 and 4·896.

10. A mass of 5 lbs. having an initial velocity of 4 feet per second is acted upon by a force of 10 lbs. in the same direction as the initial motion. What will be the velocity at the end of two seconds? *Ans.:* 132·8.

11. What is the *momentum* of a body whose mass is 3 lbs. and velocity 9 feet per second? *Ans.:* 27.

12. What momentum will be produced by a force of f pounds acting for t seconds? *Ans.:* gft.

13. If a body which weighs 9 lbs. have its velocity altered from 3 feet per second to 11 feet per second in the same straight line, what is its *change of momentum*? *Ans.:* 72.

14. What force must have acted upon a body for t seconds

to have produced a change of momentum equal to M? *Ans.:* M $\div gt$.

15. If a mass of 4 lbs. move with a velocity of 1 foot per second in a circle whose radius is 1 foot, determine the force with which the mass is urged towards the centre? *Ans.:* $4 \div g$.

16. A railway truck weighing 5 tons is moving round a curve of 100 yards radius with a velocity of 10 miles per hour. What horizontal force must be applied by the rails? *Ans.:* 249 lbs.

CHAPTER XIV.

THE THIRD LAW OF MOTION.

§ 94. **Enunciation.**—The third law of motion which was enunciated by Newton may be thus expressed:

To every action there is always an equal and contrary reaction, or the mutual actions of two bodies are always equal and oppositely directed.

§ 95. **Illustrations.**—Take first the very simple case of an object, suppose a book, resting upon a table. The book is acted upon by gravitation. Yet, as it remains at rest, the effect of gravitation must be exactly balanced by some equal and opposite force; that force is the reaction of the table. A magnet attracts a piece of iron, but at the same time, the iron also attracts the magnet with a force which is precisely equal and opposite. This may be proved experimentally by fastening a magnet in the bow of a small boat and a piece of iron in the stern. The attraction which the magnet exerts upon the piece of iron produces a force which tends to urge the boat onwards, but experiment shows that we cannot thus move the boat. No matter how powerful the magnet may be or how easily the boat may be moved, it is found to remain at rest. This can only be because the piece of iron exerts a force on the magnet which tends to pull the boat backwards with precisely the same force as the attraction of the magnet on the iron tends to pull it forwards. If the magnet be held in the hand, as in the common toy, then the reaction is sustained by the hand and not by the boat, and the action of the magnet upon a piece of iron attached to the boat, not being neutralised by a corresponding reaction, the boat will move along.

§ 96. **Atwood's Machine.** Let P and Q be two weights which are attached to the extremities of a cord that passes over a pulley and let P be greater than Q. Let X denote the action of the weight P upon the weight Q, and let Y denote the reaction of the weight Q upon P. It is to be shown that the forces X and Y are identical. The effective force which acts on P is P—Y, while the force on Q is X—Q. As the velocity of the two weights is always the same, though, of course, in opposite directions, we must have the equation.

$$\frac{P-Y}{P}=\frac{X-Q}{Q}.$$

This is, however, not sufficient, for we have only one equation between the two unknowns X and Y. We have, therefore, to resort to experiment, to obtain another equation which will render the values of X and Y determinate. It is observed that the velocity which is acquired in a given time is equal to the velocity which would have been acquired in the same time by a mass of P+Q, if acted upon by a force P−Q, whence we deduce:

$$\frac{P-Y}{P}=\frac{P-Q}{P+Q},$$

$$\frac{X-Q}{Q}=\frac{P-Q}{P+Q}.$$

from which:

$$X=Y=\frac{2PQ}{P+Q}.$$

The observed facts are, therefore, only consistent with the supposition that even during the motion of Atwood's machine the action of the weight P upon Q is precisely equal and opposite to the reaction of Q upon P. This illustration will show that while the third law of motion expresses an undoubted truth, yet like the other two laws of motion, that truth is only to be established by an induction based on observation and experiment.

LONDON: PRINTED BY
SPOTTISWOODE AND CO., NEW-STREET SQUARE
AND PARLIAMENT STREET

Printed in Dunstable, United Kingdom

91219495R00105